AF358685

Food Pesticide Residues Monitoring and Health Risk Assessment

Food Pesticide Residues Monitoring and Health Risk Assessment

Editors

Guangyang Liu
Fengnian Zhao

Basel • Beijing • Wuhan • Barcelona • Belgrade • Novi Sad • Cluj • Manchester

Editors

Guangyang Liu
Institute of Vegetables
and Flowers
Chinese Academy of
Agricultural Sciences
Beijing
China

Fengnian Zhao
School of Light Industry
Science and Engineering
Beijing Technology and
Business University
Beijing
China

Editorial Office
MDPI
St. Alban-Anlage 66
4052 Basel, Switzerland

This is a reprint of articles from the Special Issue published online in the open access journal *Foods* (ISSN 2304-8158) (available at: https://www.mdpi.com/journal/foods/special_issues/4A8FA25614).

For citation purposes, cite each article independently as indicated on the article page online and as indicated below:

Lastname, A.A.; Lastname, B.B. Article Title. *Journal Name* **Year**, *Volume Number*, Page Range.

ISBN 978-3-7258-0383-5 (Hbk)
ISBN 978-3-7258-0384-2 (PDF)
doi.org/10.3390/books978-3-7258-0384-2

Contents

Editorial

Food Pesticide Residues Monitoring and Health Risk Assessment

Yuwei Hua and Guangyang Liu *

Institute of Vegetables and Flowers, Chinese Academy of Agricultural Sciences, State Key Laboratory of Vegetable Biological Breeding, Key Laboratory of Vegetables Quality and Safety Control, Laboratory of Quality & Safety Risk Assessment for Vegetable Products (Beijing), Ministry of Agriculture and Rural Affairs of China, Beijing 100081, China; 18434762011@163.com
* Correspondence: liuguangyang@caas.cn; Tel.: +86-010-82109532

Citation: Hua, Y.; Liu, G. Food Pesticide Residues Monitoring and Health Risk Assessment. *Foods* **2024**, *13*, 474. https://doi.org/10.3390/foods13030474

Received: 21 January 2024
Accepted: 26 January 2024
Published: 2 February 2024

This Special Issue presents a share of the work published in the journal *Foods* on pesticide residue monitoring and risk assessment in food. The topics presented herein included the pre-treatment of pesticide residues in food, detection methods for pesticide residues, and the risk assessment of pesticide residues [1,2]. These works targeting these themes are summarized below.

In order to accurately detect the levels of various kinds of pesticides across different food types, effective extraction, clean-up, and enrichment of the samples are imperative. Currently, the commonly used pretreatment techniques for pesticide residues include solid-phase extraction (SPE), solid-phase microextraction (SPME), magnetic solid-phase extraction (MSPE), matrix dispersive solid-phase extraction (MDSE), QuEChERS, gel permeation chromatography, molecularly imprinted solid-phase extraction (MISPE), and microwave-assisted extraction (MEA) [3,4]. Donghui Xu et al. (contribution 1), from the Institute of Vegetable and Flower Research of the Chinese Academy of Agricultural Sciences (CAAS), constructed a nanocomposite for magnetic solid-phase extraction of pyrethroids in tea beverages. Magnetic MWCNTs-ZIF-8 was functionalized by tetrabutylammonium chloride-dodecanol (DES5) to obtain a novel magnetic nanocomposite adsorbent, which can be used for solid-phase extraction of six pyrethroids in tea. The characterization results show that MM/ZIF-8@DES5 has a high specific surface area and superparamagnetism, which is conducive to the rapid enrichment of pyrethroids in tea beverage samples. The results of their optimization experiments indicate that DES5, consisting of tetrabutylammonium chloride and 1-dodecanol, was selected for the subsequent experiments and the adsorption performance of MM/ZIF-8@DES5 was higher than that of MM/ZIF-8 and M-MWCNTs. The validation results show that the method has a wide linear range (0.5~400 µg L^{-1}, R^2 ≥ 0.9905), a low detection limit (0.08~0.33 µg L^{-1}), and good precision (intra-day RSD ≤ 5.6%, inter-day RSD ≤ 8.6%). The method was successfully applied to the determination of pyrethroid insecticides in three tea beverage samples and holds considerable promise for the monitoring of organic contaminants in environmental or food samples.

Traditional pesticide residue detection methods generally include gas chromatography–mass spectrometry (GC-MS), liquid chromatography–mass spectrometry (LC-MS), and high-performance liquid chromatography–mass spectrometry (HPLC-MS) [5]. These methods can accurately detect pesticides, but retain some limitations, such as complex and time-consuming pre-treatment steps, high training requirements for operators, expensive equipment, and inconvenience in terms of on-site analysis. In recent years, rapid detection has gradually become a popular research direction, and its application to the detection of pesticide residues can not only meet the needs of rapid, sensitive, and highly selective pesticide residue detection, but can also reduce the technical threshold for inspectors and the costs of testing [6,7]. Zhen Cao et al. (contribution 2), from the Institute of Agricultural Products Quality Standards and Testing Technology, CAAS, developed two rapid assays for the efficient determination of halosulfuron methyl (HM). High-quality anti HM monoclonal antibody (Mab, No.1A91H11) was prepared by using a pyrazosulfuronamide of HM to generate semi-antigens and antigens. A direct competitive immunoassay (dcELISA) of Mab 1A91H11 was first established to obtain a half-maximal inhibitory concentration (IC50) of 1.5×10^{-3} mg/kg, with a linear range of 0.7×10^{-3} mg/kg–10.7×10^{-3} mg/kg. The sensitivity of the assay was shown to be 10 times higher than that of indirect competitive ELISA (icELISA). The average spiked recoveries were 78.9~87.9% and 103.0~107.4%, with coefficients of variation of 1.1~6.8% and 2.7~6.4% for tomato and corn substrates spiked with 0.01, 0.05, and 0.1 mg/kg HM, respectively. In addition, a magnetic lateral flow immunoassay (MLFIA) was developed for the quantitative detection of low concentrations of HM in rice water. The sensitivity of MLFIA was 3.3~50 times higher (IC50 of 0.21×10^{-3} mg/kg) than dcELISA. The average recovery of the developed MLFIA was 81.5~92.5%, with an RSD of 5.4~9.7%. Their results show that these two methods are suitable for the rapid detection of HM residues in substrate, corn substrate, and rice water, with improved sensitivity over traditional methods. Moreover, these two methods are very practical as a rapid test that is simple and easy to conduct. Fengnian Zhao et al. (contribution 3), from the School of Chemical and Materials Engineering, Beijing Technology and Business University (BTBU), constructed a portable three-electrode sensor based on laser-induced graphene (LIG) for the electrochemical detection of carbendazim (CBZ). The LIG three-electrode sensor can easily be produced by writing directly onto the PI film using a laser. The structure and composition of LIG were verified by SEM, Raman spectroscopy, and XPS, which confirmed its highly porous graphene structure and excellent specific surface area. The sensor's detection performance was further improved by electrodepositing PtNPs onto the LIG surface. The prepared sensor (LIG/Pt) exhibited a wide linear range (1~40 μM), satisfactory LOD (0.67 μM), and good recovery (88.89~99.50%) in wastewater samples under optimal conditions. Furthermore, the electrochemical sensor is simple, sensitive, and selective, and provides a reliable real-time analytical method for the detection of CBZ residues in water samples. Fen Jin et al. (contribution 4), from the Institute of Agricultural Products Quality Standards and Testing Technology, Chinese Academy of Agricultural Sciences (CAAS), investigated the spatial distribution and migration characteristics of chloropyrifos, a plant growth regulator used for fruit pollination and fruit set, in fruit tissues. In this study, a matrix-assisted laser desorption/ionization mass spectrometry imaging (MALDI-MSI) method was developed for the first time to detect and quantify the dynamic results of clopyralid in fruits. The results show that clopyralid is mainly distributed in the pericarp and mesocarp portions of the fruit and shows a decreasing trend within 2 d of application. At the same time, the degradation rate of clopyralid was also detected using the HPLC-MS/MS method, and the results were similar to those resulting from the MALDI-MSI method, which proves that the results of detecting and quantifying clopyralid by this method are reliable. The establishment of this method provides a new means of pesticide detection and quantification, and the relevant residue results of chloropyrifuron provide data indicating the risks of its application.

Risk assessment refers to the scientific evaluation of biological, chemical, and physical hazards in food and food additives that may cause effects adverse for human health. Risk assessment is an invaluable basis for governments to formulate quality and safety standards and technical regulations for agricultural products, and it has become a necessary means for countries around the world to deal with technical barriers to trade in agricultural products and to regulate the import and export of agricultural products [8]. Zhan Dong et al. (contribution 5), from the Institute of Agricultural Products Quality Standard and Testing Technology, Shandong Academy of Agricultural Sciences, conducted the first comprehensive evaluation of the dissipation, metabolism, accumulation, handling, and risk assessment of flupyradifuran (FLU) and trichlormethrin (TRI) in cucumber and cowpea from cultivation to consumption. The results of the study show that flupyrine and TRI residues were higher in cowpea and dissipated more rapidly in cucumber. The major compounds found in the field samples were FLU and TRI ($\leq$3256.29 μg/kg), while their metabolites, FLb and TRI, fluctuated at low residue levels (≤76.17 µg/kg). In addition, FLU and TRI accumulated in both cucumber and cowpea after repeated spraying, It was shown that peeling, washing, frying, boiling, and acid washing partially or largely removed FLU and TRI residues from raw cucumber and cowpea. In contrast, residues of the metabolite TRA were significantly enriched after acid washing. The risk assessments indicated that, based on the results of this study, chronic and acute exposure to both FLU and TRI through the consumption of cucumber and cowpea poses low health risks for children or adults. In the future, more experimental sites and crop categories should be selected nationwide to study the pathways of FLU and TRI residues to get a full picture of their actual dietary risks, which would be an important step in ensuring the safe use of FLU- and TRI-containing products and protecting human health. Yaohai Zhang et al. (contribution 6), from the Institute of Citrus Research, Southwest University, China, examined 573 kumquat samples originating from China for pesticide residues using the QuEChERS, UHPLC-MS/MS, and GC-MS/MS methods, which provided data for food safety checks of kumquats and enabled the reduction of human health risks. Their results show that 90% of the samples contained one or more pesticide residues. A total of 30 pesticides were detected, including 16 insecticides, 7 fungicides, 5 acaricides, and 2 plant growth regulators. Two of the pesticides had already been banned. The frequently detected pesticides included tebuconazole, spinosad, propiconazole, cyfluthrin, isoconazole, and imidacloprid. Two or more pesticide residues were found in 81% of the samples, and 9.4% of the samples had pesticide residues exceeding the MRLs, mainly including: isopropylphosphate, bifenthrin, triazophos, avermectin, thiocyclamfen, isoconazole, and thiram. Abamectin had the highest MRL exceedance rate at 1325%. The detection rate of the pesticides in kumquats was high and multiple residues were present, with about 81% of the qualified samples being contaminated. Ji-Ho Lee et al. (contribution 7), from the Department of Crop Science, Konkuk University, studied the dissipation kinetics of spirodiclofen and phenoxypyridinium 10 d after application, using pre-harvest time intervals. Spirodiclofen and phenoxypyrazone were applied in two greenhouses in Taean-gun, Chungcheongnam-do (Daejeon 1), and Gwangyang-si, Jeollanam-do (Daejeon 2), Republic of Korea. The samples were collected at 0, 1, 3, 5, 7, and 10 d after application. The method was validated using LCMS/MS, and the spiked recoveries were 82.0~115.9%. The biological half-lives of spirodiclofen and fenpropathrin were 4.4 and 3.8 d, respectively, in field 1, and 4.5 and 4.2 d, respectively, in field 2. The pre-harvest residue limits (PHRLs) of spirodiclofen on Aster were 37.6 mg/kg (field 1) and 41.2 mg/kg (field 2), respectively, and the PHRLs of fenpropathrin on Aster were 7.2 mg/kg (Field 1) and 3.6 mg/kg (Field 2), respectively. Hazard factors at pre-harvest intervals were broadly less than 100% for both pesticides—the exception being spirodiclofen at 0 days). Moreover, the HQs of spirodiclofen >100% and >25%, at day 0 and day 7 after application, respectively, could be considered risky.

Pesticide residue digestion is affected by a combination of environmental factors, pesticide endosorption, a variety of agricultural products, cultivation methods, soil quality, and other factors. The half-life and residue amount of pesticide residues varied in different experimental regions. Since different pesticide digestion prediction models are applicable to different backgrounds and have their own advantages and disadvantages, pesticide residue modeling combined with pesticide residue detection technology plays an important role in fitting the pesticide residue digestion law. Junsong Xiao et al. (contribution 8), from the College of Food and Hygiene, Beijing Technology and Business University (BTBU), investigated the effects of temperature and relative humidity on the degradation characteristics of five pesticides (carbendazim, fensulfuron, triadimefon, chlorpyrifos, and endosulfan) in wheat and flour, and developed a quantitative prediction model. Positive samples were prepared by spraying certain concentrations of the corresponding pesticide standards, and then storing the samples at different combinations of temperature (20 °C, 30 °C, 40 °C, 50 °C) and relative humidity (50%, 60%, 70%, 80%). The samples were collected at specific time points, ground, extracted, and purified for pesticide residue detection using the QuEChERS method, and then quantified by UPLC-MS/MS. Minitab 17 software was used to model the quantification of pesticide residues. The results show that high temperature and high humidity could accelerate the degradation of five pesticide residues, and the degradation curves and half-lives of different pesticides vary with temperature and relative humidity. A quantitative model of pesticide degradation in the whole process from wheat to flour was constructed, and the R^2 of wheat and flour were greater than 0.817 and 0.796, respectively. This quantitative model can be used for the prediction of the pesticide residue levels in the process of wheat milling.

Conclusions and Outlook

In terms of the current status of pesticide residues in food, differing levels of pesticide residues were detected in these studies, some of which were below the maximum residue limit values; however, some were at exceeded permitted levels and some banned pesticides were also detected. The results of risk assessment for pesticide residues in various types of food by the relevant organizations showed that some pesticides pose a dietary intake risk. What follows is a forecast of the possible directions the development of pesticide residue testing and risk assessment may take.

Pesticide Residue Detection Methods: (1) As the target range of pesticides increases, it is necessary to fine-tune detection conditions according to the characteristics of the food, to verify and optimize the established detection parameters to ensure that the detection methods are accurate and reliable, and to promote the development of the safe and standardized use and application of pesticides. (2) The maximum pesticide residue limits should be refined scientifically and effectively on the basis of risk assessment results and taking into account the biotoxicity of the pesticides, the existing levels of residues, people's dietary intake, and other factors, to ensure that these limits are fully problem-oriented and that measures for residue management can be properly researched.

Pesticide Residue Risk Assessment: (1) Multi-methodology assessment—the currently widely adopted assessment models are limited to rough estimates of exposure and are unable to assess accumulation in target organs. Exposure assessment and early warning models can be improved to address this issue by exploring the use of relevant models from other fields. At present, many assessment methods applied in other industries have been introduced into pesticide residue risk assessment, which has led to a more diversified approach to food safety risk assessment. (2) Integrated assessment—if a particular hazardous factor in food is analyzed using only one risk assessment method, it is very vulnerable to subjectivity and methodological limitations. The risk assessment system should be brought into line with the characteristics of local populations and diets, and an assessment model should be constructed that aligns with the specific characteristics of different populations, ensuring a wider scope of application, fewer limitations, easy-to-obtain data, and greater accuracy. The simultaneous use of multiple assessment methods for comprehensive assessment will yield more comprehensive risk information.

Author Contributions: Conceptualization, Y.H. and G.L.; methodology, Y.H.; software, Y.H.; validation, Y.H. and G.L.; formal analysis, Y.H.; investigation, Y.H.; resources, G.L.; data curation, G.L.; writing—original draft preparation, Y.H.; writing—review and editing, Y.H.; visualization, Y.H. and G.L.; supervision, G.L.; project administration, G.L.; funding acquisition, G.L. All authors have read and agreed to the published version of the manuscript.

Funding: This research was funded by the National Key Research and Development Program of China (2022YFF0606800), China Agriculture Research System of MOF and MARA (CARS-23-E03).

Conflicts of Interest: The authors declare no conflict of interest.

List of Contributions:

1. Huang, X.; Liu, H.; Xu, X.; Chen, G.; Li, L.; Zhang, Y.; Liu, G.; Xu, D. Magnetic Composite Based on Carbon Nanotubes and Deep Eutectic Solvents: Preparation and Its Application for the Determination of Pyrethroids in Tea Drinks. *Foods* **2022**, *12*, 8.

2. Ying, Y.; Cui, X.; Li, H.; Pan, L.; Luo, T.; Cao, Z.; Wang, J. Development of Magnetic Lateral Flow and Direct Competitive Immunoassays for Sensitive and Specific Detection of Halosulfuron-Methyl Using a Novel Hapten and Monoclonal Antibody. *Foods* **2023**, *12*, 2764.

3. Wang, L.; Li, M.; Li, B.; Wang, M.; Zhao, H.; Zhao, F. Electrochemical Sensor Based on Laser-Induced Graphene for Carbendazim Detection in Water. *Foods* **2023**, *12*, 2277.

4. Wang, Q.; Li, X.; Wang, H.; Li, S.; Zhang, C.; Chen, X.; Dong, J.; Shao, H.; Wang, J.; Jin, F. Spatial Distribution and Migration Characteristic of Forchlorfenuron in Oriental Melon Fruit by Matrix-Assisted Laser Desorption/Ionization Mass Spectrometry Imaging. *Foods* **2023**, *12*, 2858.

5. Cui, K.; Guan, S.; Liang, J.; Fang, L.; Ding, R.; Wang, J.; Li, T.; Dong, Z.; Wu, X.; Zheng, Y. Dissipation, Metabolism, Accumulation, Processing and Risk Assessment of Fluopyram and Trifloxystrobin in Cucumbers and Cowpeas from Cultivation to Consumption. *Foods* **2023**, *12*, 2082.

6. Zhang, Y.; Li, Z.; Jiao, B.; Zhao, Q.; Wang, C.; Cui, Y.; He, Y.; Li, J. Determination, Quality, and Health Assessment of Pesticide Residues in Kumquat in China. *Foods* **2023**, *12*, 3423.

7. Saini, R.K.; Shin, Y.; Ko, R.; Kim, J.; Lee, K.; An, D.; Chang, H.-R.; Lee, J.-H. Dissipation Kinetics and Risk Assessment of Spirodiclofen and Tebufenpyrad in Aster scaber Thunb. *Foods* **2023**, *12*, 242.

8. Ding, Z.; Lin, M.; Song, X.; Wu, H.; Xiao, J. Quantitative Modeling of the Degradation of Pesticide Residues in Wheat Flour Supply Chain. *Foods* **2023**, *12*, 788.

References

1. Xu, L.; Abd El-Aty, A.M.; Eun, J.B.; Shim, J.H.; Zhao, J.; Lei, X.; Gao, S.; She, Y.; Jin, F.; Wang, J.; et al. Recent Advances in Rapid Detection Techniques for Pesticide Residue: A Review. *J. Agric. Food Chem.* **2022**, *70*, 13093–13117. [CrossRef] [PubMed]

2. Bass, C.; Denholm, I.; Williamson, M.S.; Nauen, R. The global status of insect resistance to neonicotinoid insecticides. *Pestic. Biochem. Physiol.* **2015**, *121*, 78–87. [CrossRef] [PubMed]

3. Narenderan, S.; Meyyanathan, S.; Babu, B. Review of pesticide residue analysis in fruits and vegetables. Pre-treatment, extraction and detection techniques. *Food Res. Int.* **2020**, *133*, 109141. [CrossRef] [PubMed]

4. Qin, J.; Fu, Y.; Lu, Q.; Dou, X.; Luo, J.; Yang, M. Matrix-matched monitoring ion selection strategy for improving the matrix effect and qualitative accuracy in pesticide detection based on UFLC-ESI-MS/MS: A case of Chrysanthemum. *Microchem. J.* **2020**, *160*, 105681. [CrossRef]

5. Wang, Y.; Qin, J.; Lu, Q.; Tian, J.; Ke, T.; Guo, M.; Luo, J.; Yang, M. Residue detection and correlation analysis of multiple neonicotinoid insecticides and their metabolites in edible herbs. *Food Chem. X* **2023**, *17*, 100603. [CrossRef]

6. Goh, M.S.; Lam, S.D.; Yang, Y.; Naqiuddin, M.; Addis, S.N.K.; Yong, W.T.L.; Luang-In, V.; Sonne, C.; Ma, N.L. Omics technologies used in pesticide residue detection and mitigation in crop. *J. Hazard. Mater.* **2021**, *420*, 126624. [CrossRef] [PubMed]

7. Qi, H.; Wang, Z.; Li, H.; Li, F. Directionally In Situ Self-Assembled Iridium(III)-Polyimine Complex-Encapsulated Metal–Organic Framework Two-Dimensional Nanosheet Electrode To Boost Electrochemiluminescence Sensing. *Anal. Chem.* **2023**, *95*, 12024–12031. [CrossRef]

8. Han, W.; Tian, Y.; Shen, X. Human exposure to neonicotinoid insecticides and the evaluation of their potential toxicity: An overview. *Chemosphere* **2018**, *192*, 59–65. [CrossRef] [PubMed]

 *foods*

Article

Determination, Quality, and Health Assessment of Pesticide Residues in Kumquat in China

Yaohai Zhang [1,2,3,4,5,*], Zhixia Li [1,2,3,4,5], Bining Jiao [1,2,3,4,5], Qiyang Zhao [1,2,3,4,5], Chengqiu Wang [1,2,3,4,5], Yongliang Cui [1,2,3,4,5], Yue He [1,2,3,4,5] and Jing Li [1,2,3,4,5]

[1] Citrus Research Institute, Southwest University, Chongqing 400712, China; lizhixia@cric.cn (Z.L.); jiaobining@cric.cn (B.J.); zhaoqiyang@cric.cn (Q.Z.); wangchengqiu@cric.cn (C.W.); cuiyongliang@cric.cn (Y.C.); heyue@cric.cn (Y.H.); lijing@cric.cn (J.L.)
[2] Quality Supervision and Testing Center for Citrus and Seedling, Ministry of Agriculture and Rural Affairs, Chongqing 400712, China
[3] Key Laboratory of Quality and Safety Control of Citrus Fruits, Ministry of Agriculture and Rural Affairs, Chongqing 400712, China
[4] Laboratory of Quality and Safety Risk Assessment for Citrus Products, Ministry of Agriculture and Rural Affairs, Chongqing 400712, China
[5] National Citrus Engineering Research Center, Chongqing 400712, China
[*] Correspondence: zhangyaohai@cric.cn; Tel.: +86-23-68349046

Abstract: Pesticide residues in kumquat fruits from China, and the quality and chronic/acute intake risks in Chinese consumers, were assessed using the QuEChERS procedure and UHPLC-MS/MS and GC-MS/MS methods. Our 5-year monitoring and survey showed 90% of the 573 samples of kumquat fruits collected from two main production areas contained one or multiple residual pesticides. Overall, 30 pesticides were detected, including 16 insecticides, 7 fungicides, 5 acaricides, and 2 plant growth modulators, of which 2 pesticides were already banned. Two or more residual pesticides were discovered in 81% of the samples, and pesticide residues in 9.4% of the samples surpassed the *MRLs*, such as profenofos, bifenthrin, triazophos, avermectin, spirodiclofen, difenoconazole, and methidathion. The major risk factors on the safety of kumquat fruits before 2019 were profenofos, bifenthrin, and triazophos, but their over-standard frequencies significantly declined after 2019, which was credited to the stricter supervision and management policies by local governments. Despite the high detection rates and multi-residue occurrence of pesticides in kumquat fruits, about 81% of the samples were assessed as qualified. Moreover, the accumulative chronic diet risk determined from *ADI* is very low. To better protect the health of customers, we shall formulate stricter organic phosphorus pesticide control measures and stricter use guidelines, especially for methidathion, triazophos, chlorpyrifos, and profenofos. This study provides potential data for the design of kumquat fruit quality and safety control guidelines and for the reduction in health risks to humans.

Keywords: kumquat; pesticide residue; risk assessment

Citation: Zhang, Y.; Li, Z.; Jiao, B.; Zhao, Q.; Wang, C.; Cui, Y.; He, Y.; Li, J. Determination, Quality, and Health Assessment of Pesticide Residues in Kumquat in China. *Foods* **2023**, *12*, 3423. https://doi.org/10.3390/foods12183423

Academic Editor: Roberto Romero-González

Received: 29 June 2023
Revised: 10 September 2023
Accepted: 12 September 2023
Published: 14 September 2023

1. Introduction

Citrus fruits are one of the major commodity fruits worldwide and rank first among all fruits in terms of yields. The global annual trading amount of citrus fruits ranks only after wheat and corn, which makes citrus the third-largest international commercial agricultural product worldwide [1]. Citrus fruits are among the predominant agricultural products of China, which has the largest planting area and yield in the world. There are five main varieties of citrus in China, including loose-skin mandarin, sweet oranges, pomelos, lemons, and kumquats. Kumquat, belonging to *Fortunella*, is a relative of citrus and both of them belong to Rutaceae. Kumquat originated from South Asia, being planted for over 1600 years in the Asia-Pacific and grown worldwide [2]. The main varieties of kumquat include Jindan (*F. crassifolia*), Luofu (*F. margarita*), Luowen (*F. japonica*), Jindou (*F. hindsill*), and other kumquats (*intergeneric hybrids*) according to the Records of Chinese Fruit Trees—Kumquat Fruits [3,4]. Different from most citrus fruits, whole fruits of kumquat are edible, with an intense sweet start and a slightly bitter finish. The annual production of kumquats in 2019 was about 600,000 tons in China, and the total planting area was nearly 390,000,000 m^2, making China rank first in terms of both yield and planting area. In 2022, Yangshuo and Rong'an of Guangxi accounted for 99% of total yield in China.

Kumquat fruits contain various nutrients and trace elements necessary for the human body, such as vitamins, amino acids, sugars, minerals, pectins, and dietary fibers. Most importantly, kumquat fruits have unique nutritional functional components, which are different from other citrus varieties [5–7]. The kumquat fruit planting areas of China are mainly located at the subtropics with north latitudes 22~33° and altitudes below 800 m. The climate in these areas is dominated by high temperature, rains, warmth, and wetness. Together with the long growth period of kumquat fruits and risks of diverse pests and hazards, the quality and security issues of kumquat fruits, including pesticide residues, are always the concern of governments and the public.

Quality and safety are major topics of agricultural products, and are decided by pesticide residues and other risk factors. For this reason, pesticide residues need to be monitored and relevant management measures are taken to prevent food chain pollution [8,9]. Moreover, years of monitoring results indicate that understanding the changing trends of pesticide residues will effectively guide managers in future control works. Currently, supervisory institutions and organizations in many countries provide laws and provisions to regulate and detect the use of pesticides [10]. In China, the Ministry of Agriculture and Rural Affairs is in charge of organizing and implementing pesticide residue evaluation and monitoring. To make sure whether agricultural products have risks for customers, the national security institution recently monitored the diet intake risks of pesticide residues [11–13].

The risks in dietary intake of pesticide residues were computed using the consumption and pollution data from national monitoring projects and combining these with international criteria [14]. Then, the result of intake risk was compared with the toxicological reference 74 value (usually the acceptable daily intake or acute reference dose, namely *ADI* or *ARfD*) 75 to validate whether residues in foods are in accord with the corresponding maximum residue 76 limit (*MRL*) [15]. To protect customer health and the environment, the *MRLs* of pesticide residues in foods are provided according to Chinese legislation GB 2763 [16]. So far, there is little research on pesticide residue monitoring and quality security assessment in oranges in China [17,18].

We collected 2922 samples of mandarins and oranges between 2013 and 2018, and detected pesticide residues using the QuEChERS procedure and UHPLC-MS/MS, GC-MS, and GC methods to evaluate the dietary risks to Chinese customers [17]. Results showed that the top risk factors were isocarbophos, triazophos, and carbofuran before 2015, and were gradually dominated by profenofos and bifenthrin after 2016. Chronic dietary risks are acceptable to both general adults and children and will not affect health. Moreover, we detected 16 common insecticides and acaricides in 1633 specimens of oranges (including 261 samples of kumquat fruits from nine varieties) using the QuEChERS procedure and UHPLC-MS/MS and GC-MS/MS methods to systematically analyze the potential health risks of the residues [18]. Results show the safety ranks as lemons > pomelos > ponkan > satsuma mandarin > oranges > citrus hybrids > 'nanfengmiju' mandarin > 'shatangju' mandarin > kumquat. However, triazophos in all varieties caused acute diet risks to customers, and bifenthrin in 'nanfengmiju' mandarin caused acute diet risk to children, which are both unacceptable.

As China is the world's largest kumquat fruit producer, it is necessary to determine the actual status of kumquat fruits in China for the sake of effective production, supervision, and safe consumption. This study aimed to (i) analyze pesticide residue levels in kumquat fruits of China and the temporal variations of pesticides that exceed their *MRLs*, (ii) evaluate the overall product quality of kumquat fruits in China using the index of quality for residues

(*IqR*), and (iii) assess whether the intake levels pose a long-term health risk to the local consumers.

2. Materials and Methods

2.1. Chemicals and Standards

Overall, 89 types of forbidden, severely restricted, and common pesticides in citrus production were chosen for detecting in the state-wide detection system, including acephate, acetamiprid, aldicarb, amitraz, avermectin, azoxystrobin, bifenthrin, boscalid, bromopropylate, buprofezin, carbendazim, carbofuran, carbosulfan, chlordimeform, chlorfluazuron, chlorothalonil, chlorpyrifos, clothianidin, coumaphos, cyhalothrin, cypermethrin, deltamethrin, 2,4-dichlorophenoxyacetic acid (2,4-D), dichlorvos, dicofol, difenoconazole, diflubenzuron, dimethoate, dipterex, o,p'-DDT, p,p'-DDD, p,p'-DDE, p,p'-DDT, emamectin, fenamiphos, fenitrothion, fenpropathrin, fenpyroximate, fenthion, fenvalerate, fipronil, fludioxonil, flusilazole, fonofos, forchlorfenuron, α-HCB, β-HCB, γ-HCB, δ-HCB, hexythiazox, imazalil, imibenconazole, imidacloprid, isazofos, isocarbophos, isofenphosmethyl, kresoxim-methyl, malathion, metalaxyl, methamidophos, methidathion, methomyl, monocrotophos, myclobutanil, 1-naphthaleneacetic acid, omethoate, paclobutrazol, parathion, parathion-methyl, permethrin, phenothiocarb, phenthoate, phorate, phosmet, phosphamidon, phoxim, pirimicarb, posfolan-methyl, posfolan-methyl, prochloraz, profenofos, propargite, propiconazole, pyridaben, pyrimethanil, quinalphos, spirodiclofen, sulfotep, tebuconazole, terbufos, thiabendazole, thiophanate-methyl, triadimefon, triazophos, and trifloxystrobin. Standard individual chemicals were bought from Dr. Ehrenstorfer GmbH (Augsburg, Germany) and the Environmental Quality Supervision and Testing Center, Ministry of Agriculture (Tianjin, China). HPLC-grade methanol and acetonitrile were from CNW (Augsburg, Germany). HPLC-level acetone and formic acid were from Kelong Chemcial Reagent Co. Ltd. (Chengdou, China). Anhydrous $MgSO_4$ and NaCl were at analytical grade (Sinopharm Chemcial Reagent Co. Ltd., Shanghai, China). Primary secondary amine (PSA) sorbent was from CNW (40–63 μm, 6 nm, Germany).

Stock solutions (1000 mg·L^{-1}) of individual pesticides were prepared in acetone, n-hexane, or methanol and kept in a brown glass storage bottle at $-50\,^\circ$C until used. The solutions were fully stable for about one year. Standard working solutions at 10 mg·L^{-1} were made by diluting the stock solution into acetone for GC-MS/MS and into acetonitrile for LC-MS/MS. Accordingly, matrix-matched standard solutions at 10–2000 μg·L^{-1} were made by adding blank sample extracts to each diluted standard solution. All water used here was deionized water (18 MΩ cm) from a Milli-Q Advantage A10 SP reagent water device (Millipore, MA, USA).

2.2. Apparatuses

A Shimadzu Nexis 2030 gas chromatograph equipped with a programmed split/splitless injector and an AOC 6000 multifunction autosampler (Shimadzu, Kyoto, Japan), and a Shimadzu 8040 138 NX tandem mass spectrometry (Shimadzu, Kyoto, Japan) were used to perform gas chromatography coupled with tandem mass spectrometry (GC-MS/MS) confirmation. An SH-Rxi-5Sil MS (30 m 140 $\times$ 0.25 mm id $\times$ 0.25 μm film) capillary column was used. A 1290 Infinity UHPLC system was linked to a 6495 Triple Quadrupole LC-MS/MS device added with a jet stream EI source (Agilent, Santa Clara, CA, USA). Data were acquired and analyzed on an Agilent MassHunter Workstation B.07.00. Chromatographic isolation was finished on an Agilent ZORBAX Eclipse Plus C_{18} column (50 mm $\times$ 2.1 mm, 1.8 μm) with gradient elution.

Samples were prepared using a GENIUS 3 vortex agitator (IKA, Stauffen, Germany), a CL31R multispeed refrigerated centrifuge (Thermo Scientific, Waltham, MA, USA), a WD12 water bath nitrogen blowing instrument (Aosheng Instrument, Hangzhou, China), and a CK2000 high-throughput tissue grinder (Thmorgan Biotechnology, Beijing, China).

2.3. Design of Sampling Plan

The sampling plan for kumquat fruit detection included two parts. First, fruits that will involve the major commodities on the main producing areas were sampled. Hence, kumquat fruits were mainly chosen from Yangshuo and Rong'an of Guangxi province. Second, sufficient pesticide choosing was ensured, and three sets of pesticides were considered: commonly used ones in kumquat fruit growth, newly registered ones for kumquat, and non-compliant ones with *MRLs* or prominent contributors to the Chinese dietary pesticide intake as per the detected results from the last years.

2.4. Sampling

From 2016 to 2020, 573 ripe kumquat fruit samples in total were obtained (Figure 1). The sample sources were mostly from plantations and professional cooperatives, and a few from markets as per the official directive process on sampling. Sediments were removed via homogenization before extraction. All kumquat fruits (3 kg each) were in the form of whole fruits. A typical part of the samples (200 g each) was chopped and homogenized in a food chopper. The homogenized samples were stored in sealed polyethylene bottles at $-20\,^{\circ}\text{C}$. The frozen samples were immediately moved to our laboratory using sealed containers with enough ice and kept frozen until tested within one month.

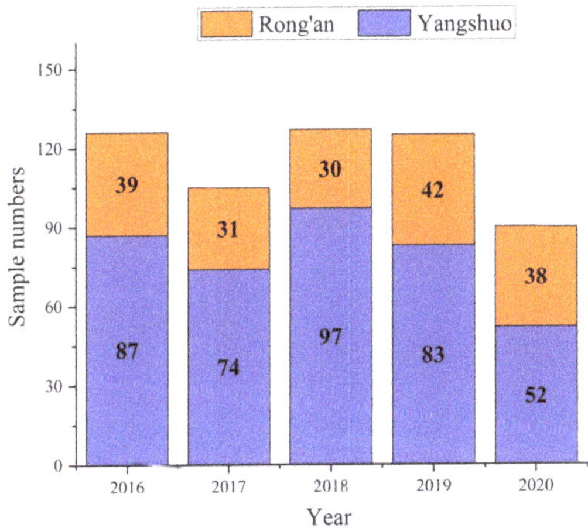

Figure 1. Number of kumquat fruit samples tested during 2016–2020.

2.5. Sample Treatment

Extraction was performed as per the QuEChERS procedure with appropriate modification and optimization. The QuEChERS procedure was elaborated below. (1) A sample (10.00 ± 0.01 g) was placed into a 50 mL FEP centrifuge tube. (2) Acetonitrile (10.00 mL) was added into each tube and oscillated heavily for 1 min. (3) The tubes were kept in a refrigerator at $-20\,^{\circ}\text{C}$ for at least 15 min. (4) Anhydrous $MgSO_4$ (4.0 g) and 1.0 g of NaCl were added, shaken heavily for 1 min, and (5) centrifuged at 10,000 rpm for 5 min. (6) Extracts (upper layer; 3.00 mL) were decanted into the centrifuge tube with 50 mg of PSA and 300 mg of anhydrous $MgSO_4$. (7) The tubes were well capped, vortexed for 1 min, and (8) centrifuged at 4000 rpm for 5 min. (9) Extracts (upper layer; 1.00 mL) were removed into a centrifuge tube, concentrated in a N_2 stream at $45\,^{\circ}\text{C}$ to dryness, redissolved in 1.00 mL of acetone, and filtered via a 0.22 μm membrane filter for GC-MS/MS. (10) The residues were filtered in the same way.

2.6. Instrumental Analysis

Fifty-two pesticides were detected using GC-MS/MS. The carrier gas was helium ($\geq$99.999%) at a flow rate of 1.0 mL/min. The injector was maintained at $250\,^{\circ}\text{C}$. Injection was performed in the pulse splitless mode. The injection volume was 1.0 µL. The column temperature program was for 60 °C at 1 min, heating first at 40 °C/min to 180 °C, and then at 8 °C/min to 280 °C, and holding for 0 min. The MS setting was as follows: data acquisition in the electron impact (EI) mode at a 70 eV voltage under the multi-reaction monitoring (MRM) mode, transmission line and ion source temperatures at 280 and 200 °C, respectively, and solvent delay time of 3 min. The MS parameters for the 52 pesticides in GC-MS/MS are shown in Table 1.

Table 1. The MS parameters for the 89 pesticides in GC-MS/MS and LC-MS/MS detection.

Analyte	Method	Quantitative Ion (*m/z*)	Qualitative Ion (*m/z*)	CE (eV)
bifenthrin	GC-MS/MS	181.1 > 166.1	181.1 > 179.1	12; 12
bromopropylate	GC-MS/MS	340.9 > 182.9	340.9 > 184.9	18; 20
buprofezin	GC-MS/MS	172.1 > 57.0	175.1 > 132.1	14; 12
chlordimeform	GC-MS/MS	196.0 > 181.0	181.0 > 140.0	10; 15
chlorothalonil	GC-MS/MS	263.9 > 168.0	263.9 > 228.8	24; 18
chlorpyrifos	GC-MS/MS	196.9 > 168.9	313.9 > 257.9	14; 14
coumaphos	GC-MS/MS	362.0 > 109.0	362.0 > 226.0	16; 14
cyhalothrin	GC-MS/MS	181.1 > 152.1	163.1 > 91.0	24; 22
cypermethrin	GC-MS/MS	163.1 > 127.1	163.1 > 91.0	6; 14
deltamethrin	GC-MS/MS	180.9 > 151.9	252.9 > 93.0	22; 20
dicofol	GC-MS/MS	139.0 > 111.0	139.0 > 75.0	16; 28
difenoconazole	GC-MS/MS	323.0 > 265.0	265.0 > 202.0	14; 20
dimethoate	GC-MS/MS	125.0 > 47.0	125.0 > 79.0	14; 8
o,p'-DDT	GC-MS/MS	235.0 > 165.0	237.0 > 165.0	24; 28
p,p'-DDD	GC-MS/MS	235.0 > 165.0	237.0 > 165.0	24; 28
p,p'-DDE	GC-MS/MS	246.0 > 176.0	317.9 > 248.0	30; 24
p,p'-DDT	GC-MS/MS	235.0 > 165.0	237.0 > 165.0	24; 28
fenamiphos	GC-MS/MS	303.1 > 195.1	288.1 > 260.1	8; 6
fenitrothion	GC-MS/MS	277.0 > 260.0	277.0 > 109.1	6; 14
fenpropathrin	GC-MS/MS	181.1 > 152.1	265.1 > 210.1	22; 12
fenpyroximate	GC-MS/MS	213.0 > 77.1	213.0 > 168.5	24; 24
fenthion	GC-MS/MS	278.0 > 109.0	278.0 > 169.0	20; 14
fenvalerate	GC-MS/MS	225.1 > 119.1	225.1 > 147.1	20; 10
fludioxonil	GC-MS/MS	248.0 > 127.0	248.0 > 154.0	26; 20
fonofos	GC-MS/MS	137.1 > 109.1	246.0 > 137.1	8; 6
α-HCB	GC-MS/MS	180.9 > 144.9	218.9 > 182.	16; 8
β-HCB	GC-MS/MS	180.9 > 144.9	218.9 > 182.	16; 8
γ-HCB	GC-MS/MS	180.9 > 144.9	218.9 > 182.	16; 8
δ-HCB	GC-MS/MS	180.9 > 144.9	218.9 > 182.	16; 8
hexythiazox	GC-MS/MS	184.0 > 149.0	156.0 > 112.0	10; 15
imazalil	GC-MS/MS	215.0 > 173.0	215.0 > 159.0	6; 6
isazofos	GC-MS/MS	257.0 > 162.0	257.0 > 119.0	8; 18
isocarbophos	GC-MS/MS	289.1 > 136.0	230.0 > 212.0	14; 10
isofenphos-methyl	GC-MS/MS	199.0 > 121.0	241.1 > 121.1	14; 22
malathion	GC-MS/MS	173.1 > 99.0	173.1 > 127.0	14; 6
methidathion	GC-MS/MS	145.0 > 85.0	145.0 > 58.0	8; 14
monocrotophos	GC-MS/MS	127.1 > 109.0	127.1 > 95.0	12; 16
omethoate	GC-MS/MS	156.0 > 110.0	110.0 > 79.0	8; 10
parathion	GC-MS/MS	139.0 > 109.0	291.1 > 109.0	8; 14
parathion-methyl	GC-MS/MS	263.0 > 109.0	125.0 > 47.0	14; 12
permethrin	GC-MS/MS	183.1 > 153.1	183.1 > 168.1	14; 14
phenothiocarb	GC-MS/MS	160.1 > 72.0	160.1 > 106.1	10; 12
phenthoate	GC-MS/MS	273.9 > 125.0	273.9 > 246.0	20; 6
phorate	GC-MS/MS	260.0 > 75.0	231.0 > 129.0	8; 24
phosmet	GC-MS/MS	160.0 > 77.0	160.0 > 133.0	24; 14
phosphamidon	GC-MS/MS	127.1 > 109.1	127.1 > 95.1	12; 18
pirimicarb	GC-MS/MS	238.1 > 166.1	166.1 > 55.0	12; 20
posfolan-methyl	GC-MS/MS	168.0 > 109.0	168.0 > 136.0	15; 15

Table 1. *Cont.*

Analyte	Method	Quantitative Ion (*m/z*)	Qualitative Ion (*m/z*)	CE (eV)
posfolan-methyl	GC-MS/MS	255.0 > 227.0	255.0 > 140.0	6; 22
profenofos	GC-MS/MS	338.9 > 268.9	336.9 > 266.9	18; 14
propargite	GC-MS/MS	135.1 > 107.1	135.1 > 77.0	16; 24
pyridaben	GC-MS/MS	147.1 > 117.1	147.1 > 132.1	22; 14
pyrimethanil	GC-MS/MS	198.1 > 183.1	198.1 > 118.1	14; 28
quinalphos	GC-MS/MS	146.1 > 118.0	146.1 > 91.0	10; 24
sulfotep	GC-MS/MS	322.0 > 202.0	322.0 > 174.0	10; 18
terbufos	GC-MS/MS	231.0 > 128.9	231.0 > 174.9	26; 14
triadimefon	GC-MS/MS	208.1 > 181.0	208.1 > 111.0	10; 22
triazophos	GC-MS/MS	161.0 > 134.0	161.0 > 106.0	8; 14
acephate	LC-MS/MS	184.0 > 143.0	184.0 > 49.0	20; 20
acetamiprid	LC-MS/MS	223.1 > 126.0	223.1 > 90.0	27; 45
aldicarb	LC-MS/MS	213.2 > 115.9	213.2 > 88.6	30; 25
amitraz	LC-MS/MS	294.2 > 163.1	294.2 > 122.1	30; 35
avermectin	LC-MS/MS	896.6 > 752.5	896.6 > 449.4	50; 55
azoxystrobin	LC-MS/MS	404.1 > 372.1	404.1 > 329.1	8; 32
boscalid	LC-MS/MS	343.0 > 307.0	343.0 > 272.0	16; 32
carbendazim	LC-MS/MS	192.1 > 160.1	192.1 > 132.1	16; 32
carbofuran	LC-MS/MS	222.1 > 165.1	222.1 > 123.1	20; 30
carbosulfan	LC-MS/MS	381.2 > 160.2	381.2 > 118.1	12; 36
chlorfluazuron	LC-MS/MS	539.9 > 383.0	539.9 > 158.0	44; 36
clothianidin	LC-MS/MS	250.0 > 169.0	250.0 > 132.0	12; 20
2,4-dichlorophenoxyacetic acid	LC-MS/MS	219.0 > 161.0	221.0 > 163.0	15; 15
dichlorvos	LC-MS/MS	221.0 > 127.0	221.0 > 109.0	27; 23
diflubenzuron	LC-MS/MS	311.0 > 158.0	311.0 > 141.0	8; 32
dipterex	LC-MS/MS	256.9 > 220.9	256.9 > 108.9	4; 15
emamectin	LC-MS/MS	886.5 > 158.0	886.5 > 82.1	40; 60
fipronil	LC-MS/MS	435.0 > 330.0	435.0 > 250.0	12; 28
flusilazole	LC-MS/MS	316.1 > 165.0	316.1 > 247.0	24; 12
forchlorfenuron	LC-MS/MS	248.1 > 129.0	248.1 > 93.0	22; 44
imibenconazole	LC-MS/MS	413.2 > 125.9	413.2 > 170.9	30; 20
imidacloprid	LC-MS/MS	256.0 > 208.9	256.0 > 175.0	12; 12
kresoxim-methyl	LC-MS/MS	314.2 > 222.1	314.2 > 267.0	10; 0
metalaxyl	LC-MS/MS	280.2 > 160.1	280.2 > 220.1	20; 10
methamidophos	LC-MS/MS	142.1 > 125.0	142.1 > 107.1	4; 3
methomyl	LC-MS/MS	163.1 > 106.0	163.1 > 88.0	4; 0
myclobutanil	LC-MS/MS	289.1 > 125.1	289.1 > 70.1	32; 16
1-naphthaleneacetic acid	LC-MS/MS	185.0 > 141.1	185.0 > 141.1	4; 4
paclobutrazol	LC-MS/MS	294.1 > 125.2	294.1 > 70.1	36; 16
phoxim	LC-MS/MS	299.0 > 129.1	299.0 > 77.1	4; 24
prochloraz	LC-MS/MS	376.0 > 265.9	376.0 > 308.0	12; 4
propiconazole	LC-MS/MS	342.1 > 159.0	342.1 > 69.1	32; 16
spirodiclofen	LC-MS/MS	411.1 > 313.0	411.1 > 71.2	5; 15
tebuconazole	LC-MS/MS	308.1 > 125.0	308.1 > 70.0	47; 40
thiabendazole	LC-MS/MS	202.0 > 175.0	202.0 > 131.0	24; 36
thiophanate-Methyl	LC-MS/MS	343.0 > 151.0	343.0 > 93.0	20; 56
trifloxystrobin	LC-MS/MS	409.1 > 145.0	409.1 > 186.0	52; 12

CE: collision energy.

The other 37 pesticides were monitored via UHPLC-MS/MS. Mobile phase A was water containing 0.1% formic acid (*v*/*v*). Mobile phase B was methanol. The gradient was started with 10% phase B, rose slowly to 90% in 0.2–6 min, from 90% to 98% in 6–9 min, from 98% to 2% in 9–12 min, and then dropped to 10%. The column was kept at 40 °C. The flow rate was 0.3 mL/min, and the injection volume was 3.0 mL. MS conditions were as follows: the EI interface with an Agilent jet stream was adopted in both negative and positive ion modes. Analysis was performed in MRM in a single run. The temperature and flow were 150 °C and 14 L/min in the drying gas, and were 375 °C and 12 L/min in the sheath gas. The nebulizer pressure was 207 kpa (30 psi). The capillary, nozzle, and fragmentor voltages were 4000, 500, and 380 V, respectively. The MS parameters for the 37 pesticides in LC-MS/MS are shown in Table 1.

2.7. Quality Control and Assurance

The pesticide residues were quantified using external standard calibration curves. The sensitivity, linearity, precision, and accuracy of the new method were verified. Linearity was acceptable when the multi-level calibration curve (5–500 µg/L) in the linear response interval of the detector for quantification exhibited correlations $r^2 > 0.99$. Sensitivity was assessed using the limit of detection (LOD) and limit of quantification (LOQ), which were computed as the lowest dose signal-to-noise (S/N) ratios of 3 and 10, respectively, by injecting spiked fruit samples. The LOD and LOQ of the method were 2–20 and 10–50 µg·kg^{-1} for the target substances. For precision and accuracy, the recovery rates of three spiked levels (0.01, 0.05, 0.2 mg·kg^{-1}) were within 70–130% and thus met the criteria. Six recovery tests were repeated at each spiked level. The relative standard deviations (RSDs) were below 10%, so the repeatability was acceptable.

2.8. Index of Quality for Residues (IqR)

IqR was computed to test how the monitored levels of multiple pesticides impacted the total quality of the samples. *IqR* for each sample was determined as the sum of the ratio of each pesticide concentration to the *MRL* (Equation (1)). The *MRLs* were cited from the Chinese National Food Safety Standard GB 2763 [16]. The citrus fruits were separated by *IqR* into 4 quality classes: Inadequate (*IqR* > 1.0), Adequate (0.6–1.0), Good (0–0.6), and Excellent (0) [19].

$$IqR = \sum_{i=1}^{n} PRCi / MRLi \tag{1}$$

where i is the given pesticide in each sample; PRC_i is the pesticide residue concentration of i and MRL_i is the MRL of pesticide i (both mg·kg^{-1}).

2.9. Dietary Risk Evaluation
2.9.1. Chronic Intake Risk

The national estimated daily intake (*NEDI*, mg·kg^{-1}·bw) and the chronic exposure risk (*%ADI*) of a pesticide were computed as follows:

$$NEDI = \frac{STMR \times Fi}{bw} \tag{2}$$

$$\%ADI = \frac{NEDI}{ADI} \times 100\% \tag{3}$$

where *STMR* (mg·kg^{-1}) is the median residue from supervised trials (herein the monitored average residues of pesticides in kumquat samples were used), F_i (kg/d) is the average fruit consumption, bw (kg) is the average body weight, and *ADI* (mg·kg^{-1} bw) is the acceptable daily intake of pesticide. The F_i of kumquats by Chinese is 1.96 g/person/d, and the bw for the general population (>1 yrs) and children aged 1–6 years is 53.23 and 16.14 kg, respectively [20]. The *ADIs* of the studied pesticides were acquired from GB 2763 [16]. The *%ADI* > 100 and <100 imply chronic risk is unacceptable and acceptable, respectively.

2.9.2. Acute Intake Risk

The international estimated short-term intake (*IESTI*, mg·kg^{-1}·bw) and acute exposure risk (*%ARfD*) were determined from Equations (4)–(7). According to the WHO, three types of the equations (Cases 1, 2a, 2b) were used for different commodities [14].

Case 1. ($Ue < 25$ g):

$$IESTI = \frac{LP \times HR}{bw} \tag{4}$$

Case 2a. (25 g $\leq Ue < LP$):

$$IESTI = \frac{Ue \times HR \times v + (LP - Ue) \times HR}{bw} \tag{5}$$

Case 2b. ($Ue > LP$)

$$IESTI = \frac{LP \times HR \times v}{bw} \tag{6}$$

$$\%ARfD = \frac{IESTI}{ARfD} \times 100\% \tag{7}$$

where LP (kg) is the large portion, HR (mg·kg^{-1}) is the highest residue in samples, Ue (kg) is the unit weight of the edible portion, and v is variability. Cases 1, 2a, and 2b were used for kumquat; orange and mandarin cultivars; lemon and pummelo, respectively. The data of LP, Ue and v for each commodity are listed in Ref. [18]. The $ARfDs$ of the tested pesticides were cited from the WHO database [21]. Similarly, $\%ARfD < 100$ and $\%ARfD > 100$ reflect acceptable and unacceptable acute risk, respectively.

3. Results

3.1. Detection of Pesticide Residues

The results of pesticide residues detected with UHPLC-MS/MS and GC-MS/MS are listed in Table 2. Among the 573 samples of kumquat fruits, 30 of the 89 targeted pesticides were accumulatively detected. For each sample, the result of pesticide residues was the average value of three repeated measurements. As per Chinese national standards, GB/T 6379.1 [22] and GB/T 6379.2 [23], precise data were obtained using UHPLC-MS/MS and GC-MS/MS. The reproducibility and repeatability of these methods were determined at 95% reliability. Based on the whole data, 16 insecticides (53.3%), 7 fungicides (23.3%), 5 acaricides (16.7%), and 2 plant growth modulators (6.7%) were identified. The insecticides mainly covered three types: organic phosphorus, pyrethroids, and nicotines (75% together). These pesticides are widely used in kumquat planting in China to prevent and control severe plant diseases such as citrus anthracnose, citrus canker, citrus scab, citrus melanose, and insect pests such as red spiders, bed bugs, stink bugs, scale insects, leaf miners, leafroller moths, and beetles [24,25].

The highest detection rates were found for tebuconazole and spirodiclofen (both 37.9%), followed by profenofos (35.1%), cyhalothrin (32.1%), difenoconazole (25.1%), imidacloprid (24.8%), thiophanate-methyl (24.1%), chlorpyrifos (21.3%), prochloraz (17.8%), propargite (16.9%), carbendazol (15.4%), and hexythiazox (15.2%). These data basically accord with other studies. For instance, chlorpyrifos and carbendazol are the most commonly identified [19,26,27], and the detection rates of prochloraz, tebuconazole, and acetamiprid are very high [28,29]. The high identification rates of spirodiclofen, profenofos, propargite, and hexythiazox (15–38%) are due to the common occurrence of red spiders in kumquat fruits.

In all the samples (N = 573), 9.8% of the kumquat samples were found with no pesticide residues, and 90.2% of the samples had at least one (of the 30 identified pesticides) that exceeded the quantitative limits. The concentrations of the 30 detected residual pesticides ranged from 0.01 to 2.24 mg·kg^{-1}. The pesticides at high concentrations (mg·kg^{-1}) were propargite (2.25), profenofos (2.10), thiophanate-methyl (1.49), prochloraz (1.10), difenoconazole (1.09), and triazophos (1.01). These pesticide residues were mostly fungicides and insecticides, followed by acaricides. The high concentrations may be ascribed to the wide use before and after harvesting [24]. Nevertheless, the high residue concentrations of these pesticides do not all exceed the *MRLs* of China [16]. The *MRLs* restrict the types and concentrations of pesticides in oranges, indicating pesticides were applied basically in accordance with Good Agricultural Practices.

Table 2. Occurrence of pesticide residues in kumquat fruits of China.

Pesticide	Type	No. (%) of Positive Samples	Concentration Range (mg·kg⁻¹)	Mean Valve (mg·kg⁻¹)	No. (%) of Exceedance	MRL (mg·kg⁻¹)
2,4-D	P	67 (11.7)	0.010–0.096	0.033		1
acetamiprid	I	69 (12.0)	0.010–0.194	0.042		2
avermectin	I	4 (0.7)	0.016–0.132	0.058	4 (0.70)	0.01
azoxystrobin	F	61 (10.6)	0.011–0.751	0.098		1
bifenthrin	I	35 (6.1)	0.011–0.324	0.051	7 (1.22)	0.05
buprofezin	I	43 (7.5)	0.010–0.538	0.087		1
carbendazim	F	88 (15.4)	0.010–0.665	0.068		5
carbofuran	I	2 (0.3)	0.011–0.015	0.013		0.02
carbosulfan	I	5 (0.9)	0.011–0.041	0.023		1
chlorpyrifos	I	122 (21.3)	0.010–0.340	0.051		1
cyhalothrin	I	184 (32.1)	0.010–0.185	0.052		0.2
cypermethrin	I	55 (9.6)	0.011–0.273	0.042		1
difenoconazole	F	144 (25.1)	0.010–1.092	0.077	1 (0.17)	0.6
fenpermethrin	I	49 (8.6)	0.011–0.474	0.093		5
fenpyroximate	A	11 (1.9)	0.018–0.256	0.086		0.5
fenvalerate	I	8 (1.4)	0.011–0.183	0.055		0.2
hexythiazox	A	87 (15.2)	0.010–0.055	0.020		0.5
imidacloprid	I	142 (24.8)	0.010–0.423	0.040		1
malathion	I	24 (4.2)	0.012–0.416	0.071		2
methidathion	I	3 (0.5)	0.010–0.420	0.160	1 (0.17)	0.05
paclobutrazol	P	2 (0.3)	0.013–0.116	0.064		0.5
prochloraz	F	102 (17.8)	0.011–1.104	0.099		10
profenofos	I	201 (35.1)	0.011–2.102	0.119	30 (5.24)	0.2
propargite	A	97 (16.9)	0.010–2.248	0.257		5
pyridaben	A	20 (3.5)	0.010–0.332	0.045		2
spirodiclofen	A	217 (37.9)	0.010–0.554	0.075	3 (0.52)	0.4
tebuconazole	F	217 (37.9)	0.010–0.570	0.101		2
thiophanate-methyl	F	138 (24.1)	0.010–1.493	0.178		3
triadimefon	F	14 (2.4)	0.066–0.617	0.254		1
triazophos	I	37 (6.5)	0.011–1.011	0.115	4 (0.70)	0.2

3.2. Pesticide Residues Over-Standardness, and Detection of Banned and Restricted Pesticides

The residue levels of 7 pesticides in the kumquat samples in the 5 tested years surpassed the Chinese *MRLs* [16]. The order ranked by over-standard rate is profenofos (5.24%) > bifenthrin (1.22%) > triazophos (0.70%) > avermectin (0.70%) > spirodiclofen (0.52%) > difenoconazole (0.17%) > methidathion (0.17%). The disqualified pesticides are all insecticides (except for spirodiclofen and difenoconazole), which may be related to the massive use of insecticides to control the frequently occurring insect pests in citrus trees. The detection rates of the two banned or restricted pesticides (methidathion and carbofuran) were 0.5% and 0.3%, respectively. The *MRL* is not a toxicological limit but shall be toxicologically acceptable. Exceeding the *MRL* and the use of forbidden pesticides are both representative of violating GAP. Notably, registration of methidathion to be used in kumquat fruit trees and other vegetables or fruits was canceled by the Ministry of Agriculture of China in 2015 because of its high toxicity. The MRL of methidathion in kumquat fruits was lowered from 2 [30] to 0.05 mg·kg⁻¹ [16].

Figure 2 shows the single-residue concentrations of the 30 identified pesticides and 7 over-standard pesticides, which were distributed in 45 of the kumquat samples, in which the residues of at least one pesticide exceeded the *MRL* (accounting for 7.8% of all samples). Avermectin was the most over-standard (1325%*MRL*), followed by profenofos (1051%*MRL*), methidathion (839%*MRL*), bifenthrin (648%*MRL*), triazophos (506%*MRL*), difenoconazole (182%*MRL*), and spirodiclofen (139%*MRL*). In our previous study, the over-standard rate of pear samples collected from China was 2.6%, and the over-standard rates of cyfluthrin, difenoconazole, omethoate, profenofos, pyrimethanil, and tebuconazole were 123–332% [31]. The over-standard rate of peach samples from China was 3.2%, and the over-standard rates of carbendazol, cyhalothrin, cypermethrin, deltamethrin, difenoconazole, fenbuconazole, flusilazole, and isazofos were 104–345% [13]. Among the samples of mandarins and oranges from China, the over-standard rate was 3.8%, and the over-standard rates of bifenthrin, profenofos, fenpyroximate, carbofuran, triazophos, isocarbophos, difenoconazole, and cyhalothrin were 186–283% [17]. Reports show pesticide residues in fruit samples exceed the *MRLs* in other countries or organizations. For instance, Mac Loughlin et al. found that in the 135 samples of fruits and vegetables in Argentina markets, the largest residual pesticide content was detected in oranges—30% of the tested oranges exceeded *MRLs* [19]. In 11% of the tested orange samples in Mexico, the concentrations of methyl chlorpyrifos, malathion, and methidathion were all over the *MRLs* of the European Union [32].

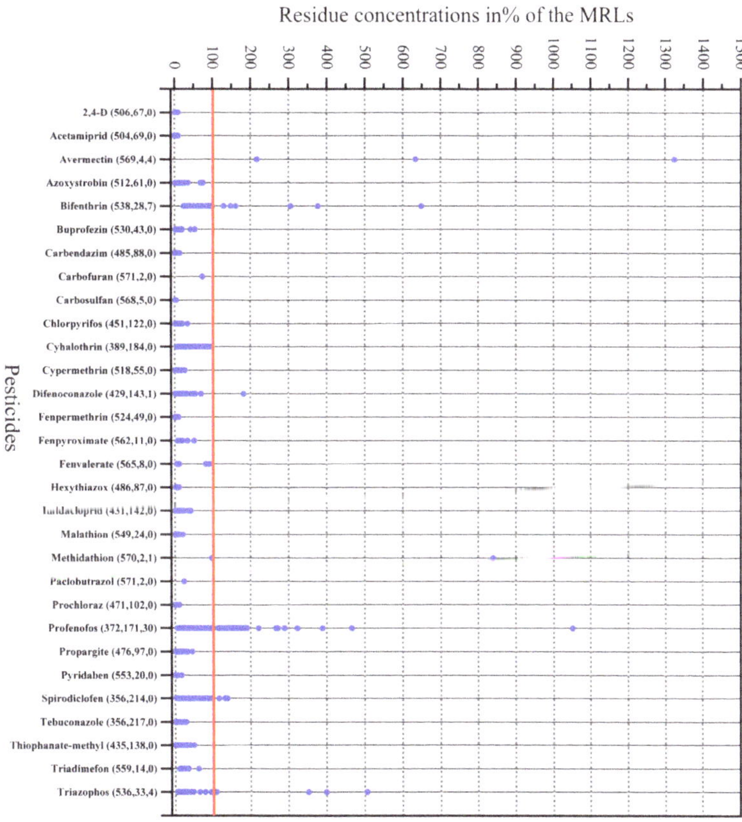

Figure 2. Residue contents of the thirty detected pesticides in kumquat samples, expressed as percentage of the *MRL* (numbers in brackets after the pesticide name refer to the numbers of samples below the LOQ, above the LOQ, below the *MRL*, and above the *MRL*. The blue dot represents the distribution of pesticide residue content and the red line represents MRL).

3.3. Multi-Residue Pesticide Residues

In the positive samples, 54 samples (9.4%) were found with one pesticide, and 463 samples (80.8%) had two or more pesticides. One sample was found with up to 12 pesticides. The detection rate changing with the number of residual pesticides maximized at four, and then gradually decreased (Figure 3). Multi-residues are ubiquitous in kumquat fruits and other fruits of many countries. Reports from the European Food Safety Authority showed that multiple pesticides were discovered in 58.7% of orange specimens. The largest number of residual pesticides (12) was found in a third country, but some oranges in the European Union contained up to 11 residual pesticides [33]. Poulsen et al. found oranges more frequently contained multiple residues than 20 other types of fruits, and multiple pesticides were detected in 75% of samples, including 93% of mandarin samples, 86% of grapefruit samples, 82% of orange samples, 79% of lemon samples, and 61% of pomelo samples [34]. When multiple pesticides are applied to treat different plant diseases and insect pests, fruits are more susceptible to multi-pesticide residual pollution.

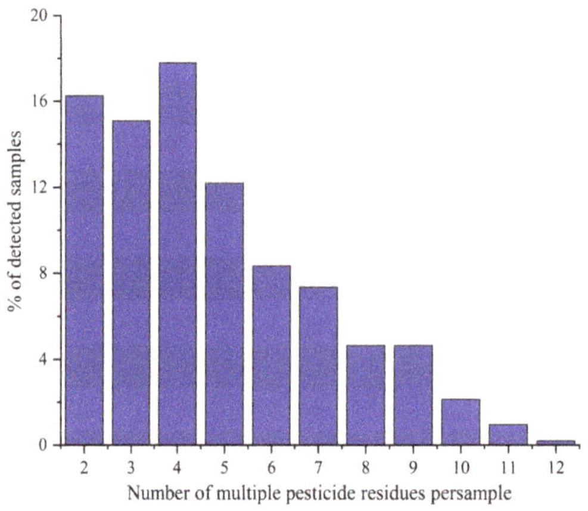

Figure 3. Proportion of multi-pesticide residues in kumquat fruit samples.

Though the amounts and types of multi-residues vary along with the planting sites, years, and planters, general rules have been discovered. Sampling and testing of fruits and vegetables in markets of Argentina showed chlorpyrifos, endosulfan, and at least one pyrethroid and one fungicide (chlorpyrifos + endosulfan + pyrethroid + fungicide) co-existed in the same sample. The most common pesticide residue combinations in orange samples were acetamiprid + chlorpyrifos + prochloraz + carbendazol (oranges); chlorpyrifos + prochloraz + etoxazole + tebuconazole, and profenofos + acetamiprid + chlorpyrifos + spirotetramat (mandarins) [17]. Moreover, multi-residue pesticides included tebuconazole + spirodiclofen + thiophanate-methyl + prochloraz, and spirodiclofen + profenofos + cyhalothrin + imidacloprid.

3.4. Changes of Over-Standard Pesticide Residues in Five Years

The temporal changes of over-standard frequency in seven over-standard residual pesticides are shown in Figure 4, including profenofos (Figure 4a), bifenthrin (Figure 4b), triazophos (Figure 4c), avermectin (Figure 4d), spirodiclofen (Figure 4e), difenoconazole (Figure 4f), and methidathion (Figure 4g). The major risk factors affecting the safety of kumquat fruits before 2019 were profenofos, bifenthrin, and triazophos. After 2019, however, the over-standard frequencies of profenofos, bifenthrin, and triazophos significantly declined, and the banned or restricted pesticides including methidathion were not over-standard, which was credited to the stricter supervision and management policies by local governments. During 2016–2018, the over-standard frequency of profenofos in kumquat fruits slightly increased, which is basically consistent with a previous report that the over-standard frequencies of profenofos in samples of mandarins and oranges steadily increased year by year [17]. Profenofos is easily over-standard in kumquats, which can be explained by three reasons. First, profenofos is registered in only a few products of oranges, and can efficiently prevent and cure red spiders. Currently, profenofos products that are not registered in oranges must have been extensively and illegally used in oranges. Second, farmers use pesticides irregularly, and may increase application doses and times as well as using pesticides at the late mature stage. Third, the residue decomposition rate of profenofos in oranges is associated with the variety and producing environment. Profenofos is a slowly degrading pesticide with a safe period over 60 days. Thus, reasonable use of profenofos and popularization of its substitutes is of concern. As high-risk pesticides are effectively controlled, the over-standard rate of kumquat fruits significantly drop year after year, and the over-standard pesticides are also random. The unqualified rate of kumquat fruits in 2020 slightly rose from that in 2019. One main reason was that GB 2763-2020 modified the limits of some pesticides used to kumquat fruits, which was a 'decreased dose' for most pesticides [35]. Thus, irregular pesticide use will increase the quality and safety risks of kumquat fruits. For instance, the limit of avermectin in kumquat fruits provided in GB 2763-2020 dropped from 0.02 to 0.01 mg·kg^{-1}, and the limit of spirodiclofen decreased from 0.5 to 0.4 mg·kg^{-1} [35]. Thus, the changes in the limits of pesticides in kumquat fruits provided in GB 2763-2020 shall be further popularized so as to normalize the application of spirodiclofen, avermectin, and other pesticides.

3.5. Quality Assessment of Kumquat Fruits

The quality safety of agricultural products in terms of pesticide residues is assessed using *MRLs* worldwide. The multiple residues existing in a single specimen will impact the quality of the whole product through the accumulation or synergistic effect of single residues. *IqR* is an effective indicator to measure the overall quality of foods [24]. In the present study, the quality in the majority of samples is satisfactory (Table 3). Clearly, 9.8%, 58.8%, and 12.4% of the samples were rated as excellent (*IqR* = 0), good (0–0.6), and adequate (0.6–1.0), respectively. The remaining 19.0% of the specimens were inadequate (*IqR* > 1.0), but this proportion is obviously higher than those reported in mandarins and oranges from China [17]. Because of the standardization of relevant data, *IqR* allows for simple and objective comparison. The above data indicate future orchard gardeners can formulate more targeted schemes to modulate the quality of orange fruits in these producing areas. Among the unqualified categories, the pesticide residues in 45 samples (41.3%) exceeded the *MRLs*, and the accumulative pesticide residues of 64 samples (58.7%) reached or were lower than the *MRLs*. Of the 109 unqualified samples, the *IqR* of 71 samples (65.1%) varied between 1 and 2. In other words, the decline in product quality at the accumulative level exceeded 1 to 2 times of the *MRL*. The *IqR* of 34/109 (31.2%) samples was between 2 and 5, and the *IqR* of 4 (3.7%) samples exceeded 5. The *MRL* of the least qualified sample was 22.8 times the appropriate level, and is higher than the reported level [17]. Admittedly, the huge differences among varieties also make the results incomparable. Nevertheless, the accumulative risks of the pesticide residues below the *MRLs* shall also be concerned, as they largely contribute to *IqR*.

Table 3. Quality evaluation of the analyzed kumquat fruits according to the calculated *IqR* factor.

IqR	No. of Samples	%	Quality Categories
0	56	9.8	excellent
0–0.6	337	58.8	good
0.6–1	71	12.4	adequate
>1	109	19.0	inadequate

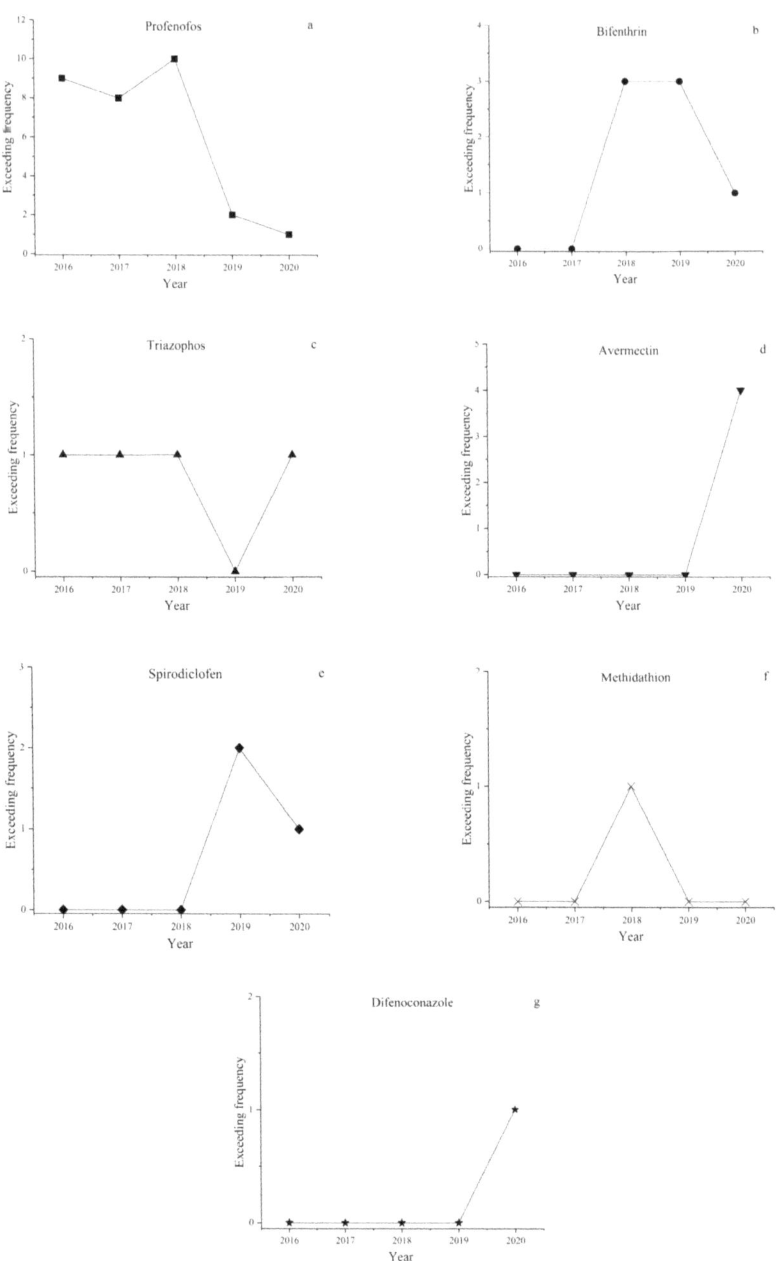

Figure 4. Temporal changes of *MRL* surpassing frequency in the seven pesticides from 2016 to 2020 ((**a**): profenofos; (**b**): bifenthrin; (**c**): triazophos; (**d**): avermectin; (**e**): spirodiclofen; (**f**): difenoconazole; (**g**): methidathion).

3.6. Health Risks of Pesticide Residues

The *%ADI* and *%ARfD* were obtained as shown in the Section 2.9. As for accumulative chronic risk assessment, the total of *%ADI* is 0.76 among general people (>1 year old) and 2.50 in children (1–6 years old), which are both smaller than 100 (Table 4). The exposure value of each pesticide is significantly below *ADI*, and the *%ADI* of each pesticide is far below 100. These data indicate the chronic risk of exposure to pesticides via eating kumquat fruits can be ignored. The *%ADI* maximized in methidathion (0.59 in children, 0.18 in general people), followed by avermectin (0.52 and 0.16, respectively) and triazophos (0.45 in children, 0.14 in general people). Generally, among the 30 detected pesticides, the residues of five organic phosphorus pesticides contributed most largely to *%ADI* (44.3%). In particular, the top four organic phosphorus pesticides were methidathion, triazophos, chlorpyrifos, and profenofos, and their total contribution was 99.9% of all organic phosphorus pesticides. The contribution rates of the other 4 insecticides (avermectin, carbofuran, bifenthrin, and buprofezin), the other 8 insecticides, 7 fungicides, and 7 pesticides were 30.5%, 4.3%, 10.4%, and 10.5%, respectively (Figure 5). Similar conclusions were made in other studies from China, Poland, and Brazil that eating fruits will not cause health risks to adults or children [17,36,37]. As for acute risk assessment, the exposure values of all pesticides are significantly lower than *ARfD* except for triazophos, and the *%ARfD* of each pesticide is far lower than 100. The *%ARfD* is the highest for triazophos (212.98 in children, 227.92 in general people), which makes up to an 89% contribution to the total of *%ARfD*. According to the above results, the use of pesticides including methidathion and carbofuran shall be more strictly controlled and punished. More importantly, high-risk organic phosphorus pesticides including triazophos, chlorpyrifos, and profenofos shall be controlled with appropriate measures or lower limits, or be completely forbidden.

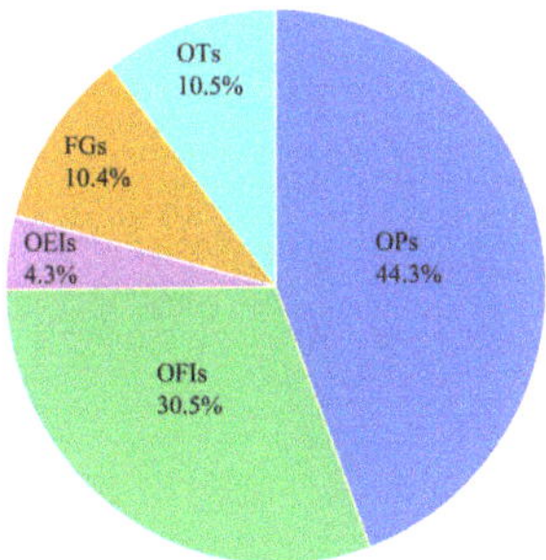

Figure 5. Contributions of the five classes of 30 detected pesticides to hazard index. OPs = organophosphorus pesticides, OFIs = other four insecticide pesticides, OEIs = other eight insecticide pesticides, FGs = fungicide pesticides, OTs = other pesticides.

The uncertainty of diet risk assessment often originates from toxicological or consumptive data, processing factors, left censored data processing, and loss of exposure assessment model [27]. Jensen et al. found the intake exposure of pesticides by humans was changeable, and was mainly decided by degradation in crops, harvesting time, and processing [38]. Clearly, processing factors are practically uncertain in assessing diet risks of primary agricultural products. Generally, pesticide residues of fruits and vegetables can be lowered by washing, soaking, peeling, blanching, or other domestic processing [39]. Moreover, the diet exposure to pesticides is reduced by 3% to 11.5% after washing, machinery, or thermal processing [40]. Here, we ignored the processing factors during diet risk assessment, so the results of exposure may be overestimated. Thus, more precise and real exposure estimation shall be made in the future.

Table 4. Health hazard index for pesticide residues in kumquat fruits.

Pesticide	ADI (mg·kg⁻¹ bw)	NEDI (mg·kg⁻¹ bw)		%ADI		ARfD (mg·kg⁻¹ bw)	IESTI (mg·kg⁻¹ bw)		%RfD	
		Gen Pop, >1 yrs	Children, 1–6 yrs	Gen Pop, >1 yrs	Children, 1–6 yrs		Gen Pop, >1 yrs	Children, 1–6 yrs	Gen Pop, >1 yrs	Children, 1–6 yrs
Chlorpyrifos	0.01	1.26×10^{-6}	4.15×10^{-6}	0.0126	0.0415	0.1	7.67×10^{-4}	7.16×10^{-4}	0.77	0.72
Malathion	0.3	1.24×10^{-6}	4.10×10^{-6}	0.0004	0.0014	2	9.37×10^{-4}	8.76×10^{-4}	0.05	0.04
Triazophos	0.001	1.36×10^{-6}	4.49×10^{-6}	0.1362	0.4493	0.001	2.28×10^{-3}	2.13×10^{-3}	227.92	212.98
Profenofos	0.03	2.11×10^{-6}	6.96×10^{-6}	0.0070	0.0232	1	4.74×10^{-3}	4.43×10^{-3}	0.47	0.44
Methidathion	0.001	1.79×10^{-6}	5.91×10^{-6}	0.1791	0.5908	0.01	9.46×10^{-4}	8.84×10^{-4}	9.46	8.84
Triadimefon	0.03	6.96×10^{-6}	2.30×10^{-5}	0.0232	0.0765	0.08	1.39×10^{-3}	1.30×10^{-3}	1.74	1.62
Cypermethrin	0.02	8.83×10^{-7}	2.91×10^{-6}	0.0044	0.0146	0.04	6.16×10^{-4}	5.76×10^{-4}	1.54	1.44
Fenvalerate	0.02	6.17×10^{-7}	2.03×10^{-6}	0.0031	0.0102	-	4.13×10^{-4}	3.86×10^{-4}	-	-
Bifenthrin	0.01	1.37×10^{-6}	4.51×10^{-6}	0.0137	0.0451	-	7.30×10^{-4}	6.83×10^{-4}	-	-
Cyhalothrin	0.02	1.29×10^{-6}	4.24×10^{-6}	0.0064	0.0212	0.02	4.18×10^{-4}	3.90×10^{-4}	2.09	1.95
Fenpermethrin	0.03	2.38×10^{-6}	7.86×10^{-6}	0.0079	0.0262	-	1.07×10^{-3}	9.99×10^{-4}	-	-
Propargite	0.01	1.47×10^{-6}	4.83×10^{-6}	0.0147	0.0483	-	5.07×10^{-3}	4.74×10^{-3}	-	-
Pyridaben	0.01	8.34×10^{-7}	2.75×10^{-6}	0.0083	0.0275	-	7.49×10^{-4}	6.99×10^{-4}	-	-
Difenoconazole	0.01	1.78×10^{-6}	5.87×10^{-6}	0.0178	0.0587	0.3	2.46×10^{-3}	2.30×10^{-3}	0.82	0.77
Fenpyroximate	0.01	2.57×10^{-6}	8.47×10^{-6}	0.0257	0.0847	0.02	5.77×10^{-4}	5.39×10^{-4}	2.89	2.70
Hexythiazox	0.03	6.33×10^{-7}	2.09×10^{-6}	0.0021	0.0070	-	1.25×10^{-4}	1.16×10^{-4}	-	-
Buprofezin	0.009	1.19×10^{-6}	3.92×10^{-6}	0.0132	0.0436	0.5	1.21×10^{-3}	1.13×10^{-3}	0.24	0.23
Tebuconazole	0.03	1.82×10^{-6}	5.99×10^{-6}	0.0061	0.0200	-	1.28×10^{-3}	1.20×10^{-3}	-	-
Acetamiprid	0.07	9.43×10^{-7}	3.11×10^{-6}	0.0013	0.0044	-	4.38×10^{-4}	4.09×10^{-4}	-	-
Carbendazim	0.03	1.35×10^{-6}	4.44×10^{-6}	0.0045	0.0148	0.5	1.50×10^{-3}	1.40×10^{-3}	0.30	-
Imidacloprid	0.06	1.01×10^{-6}	3.35×10^{-6}	0.0017	0.0056	0.4	9.53×10^{-4}	8.91×10^{-4}	0.24	0.22
Thiophanate-methyl	0.09	3.12×10^{-6}	1.03×10^{-5}	0.0035	0.0114	-	3.37×10^{-3}	3.15×10^{-3}	-	-
Spirodiclofen	0.01	1.58×10^{-6}	5.20×10^{-6}	0.0158	0.0520	-	1.25×10^{-3}	1.17×10^{-3}	-	-
Prochloraz	0.01	2.27×10^{-6}	7.48×10^{-6}	0.0227	0.0748	0.1	2.49×10^{-3}	2.32×10^{-3}	2.49	2.32
Azoxystrobin	0.2	2.36×10^{-6}	7.80×10^{-6}	0.0012	0.0039	-	1.69×10^{-3}	1.58×10^{-3}	-	-
2,4-D	0.01	9.91×10^{-7}	3.27×10^{-6}	0.0099	0.0327	-	2.17×10^{-4}	2.03×10^{-4}	-	-
Carbofuran	0.001	4.79×10^{-7}	1.58×10^{-6}	0.0479	0.1579	0.001	3.31×10^{-5}	3.10×10^{-5}	3.31	3.10
Carbosulfan	0.01	7.88×10^{-7}	2.60×10^{-6}	0.0079	0.0260	0.02	9.33×10^{-5}	8.72×10^{-5}	0.47	0.44
Avermectin	0.001	1.56×10^{-6}	5.16×10^{-6}	0.1564	0.5157	-	2.99×10^{-4}	2.79×10^{-4}	-	-
Paclobutrazol	0.1	2.37×10^{-6}	7.82×10^{-6}	0.0024	0.0078	0.05	2.62×10^{-4}	2.45×10^{-4}	0.52	0.49

Gen pop: General population.

4. Conclusions

Our 5-year monitoring and survey showed 90% of the 573 samples of kumquat fruits collected from two main production areas contained one or multiple residual pesticides. Overall, 30 pesticides were detected, including 16 insecticides, 7 fungicides, 5 acaricides, and 2 plant growth modulators, of which 2 pesticides were already banned. The commonly detected pesticides include tebuconazole, spirodiclofen, profenofos, cyhalothrin, difenoconazole, and imidacloprid. Two or more residual pesticides were discovered in 81% of the specimens, and pesticide residues surpassed the *MRLs* in 9.4% of the samples, including profenofos, bifenthrin, triazophos, avermectin, spirodiclofen, difenoconazole, and methidathion. The largest over-standard rate of 1325% *MRL* was found in avermectin. Profenofos, bifenthrin, and triazophos were the main safety risk factors of kumquat fruits before 2019, but their over-standard frequencies significantly declined after 2019, indicating the over-standard pesticides were random to some extent. Despite the high detection rates and multi-residue occurrence of pesticide residues in kumquat fruits, about 81% of the samples were assessed as qualified. Moreover, the accumulative chronic diet risk determined from *ADI* is very low. To better protect the health of customers, we shall formulate stricter organic phosphorus pesticide control measures and stricter use guidelines, especially for methidathion, triazophos, chlorpyrifos, and profenofos. This study provides potential data for the design of kumquat fruit quality and safety control guidelines and for the reduction in health risks to humans. In addition, our study performed in this manuscript improves on others performed for citrus fruits by the same authors [17,18].

Author Contributions: Y.Z., sample treatment, methodology, validation, instrumental analysis and drafting. Z.L., data analysis and paper revision. B.J., fund declaration. Q.Z., instrumental analysis. C.W., Y.C., Y.H. and J.L., design of sampling plan and sample collection. All authors have read and agreed to the published version of the manuscript.

Funding: China Agriculture Research System (No. CARS-26).

Data Availability Statement: The data used to support the findings of this study can be made available by the corresponding author upon request.

Conflicts of Interest: The authors declare no conflict of interest.

References

1. Wu, H.J. The situation and outlook of citrus processing industry in China. In Proceedings of the China/FAO Citrus Symposium, Beijing, China, 14–17 May 2001.
2. Xu, J.G.; Lin, D.S. A Study on the history of Ningbo Jingan's eastern expedition to Japan. *Chin. Agric. Hist.* **1999**, *18*, 97–101.
3. Lin, D.S.; Wu, F.C. Distribution and species of kumquat in China. *Chin. Citrus* **1987**, 3–5. [CrossRef]
4. Wang, Y.C. Research of Kumquat Cultivation History, Cultivars Resources and the Present Situation of the Industry Development of Rongan Kumquat in Guangxi Province. Master's Thesis, Guangxi University, Nanning, China, 2016.
5. Ogawa, K.; Kawasaki, A.; Omura, M.; Yoshida, T. 3′,5′-Di-C-β-glucopyranosylphloretin, a flavonoid characteristic of the genus *Fortunella*. *Phytochemistry* **2011**, *57*, 737–742. [CrossRef]
6. Barreca, D.; Bellocco, E.; Caristi, C.; Leuzzi, U.; Gattuso, G. Kumquat (*Fortunella japonica Swingle*) juice: Flavonoid distribution and antioxidant properties. *Food Res. Int.* **2011**, *44*, 2190–2197. [CrossRef]
7. Roowi, S.; Crozier, A. Flavonoids in tropical citrus species. *J. Agric. Food Chem.* **2011**, *59*, 12217–12225. [CrossRef]
8. Osman, K.A.; Al-Humaid, A.M.; Al-Rehiayani, S.M.; Al-Redhaiman, K.N. Monitoring of pesticide residues in vegetables marketed in Al-Qassim region, Saudi Arabia. *Ecotoxicol. Environ. Saf.* **2010**, *73*, 1433–1439. [CrossRef]
9. Melo, A.; Cunha, S.C.; Mansilha, C.; Aguiar, A.; Pinho, O.; Ferreira, I.M.P.L.V.O. Monitoring pesticide residues in greenhouse tomato by combining acetonitrile based extraction with dispersive liquid-liquid microextraction followed by gas chromatography-mass spectrometry. *Food Chem.* **2012**, *135*, 1071–1077. [CrossRef]
10. Islam, M.N.; Bint, E.; Naser, S.F.; Khan, M.S. Pesticide food laws and regulations. In *Pesticide Residue in Foods: Sources, Management, and Control*; Khan, M.S., Rahman, M.S., Eds.; Springer International Publishing: Cham, Swizerland, 2017; pp. 37–51.
11. Duan, Y.; Guan, N.; Li, P.P.; Li, J.G.; Luo, J.H. Monitoring and dietary exposure assessment of pesticide residues in cowpea (*Vigna unguiculata* L. Walp) in Hainan, China. *Food Control* **2016**, *59*, 250–255. [CrossRef]
12. Liu, Y.H.; Shen, D.Y.; Li, S.L.; Ni, Z.L.; Ding, M.; Ye, C.F.; Tang, F.B. Residue levels and risk assessment of pesticides in nuts of China. *Chemosphere* **2016**, *144*, 645–651. [CrossRef]

13. Li, Z.X.; Nie, J.Y.; Yan, Z.; Cheng, Y.; Lan, F.; Huang, Y.N.; Chen, Q.S.; Zhao, X.B.; Li, A. A monitoring survey and dietary risk assessment for pesticide residues on peaches in China. *Regul. Toxicol. Pharmacol.* **2018**, *97*, 152–162. [CrossRef]

14. WHO. Consultations and Workshops: Dietary Exposure Assessment of Chemicals in Food: Report of a Joint FAO/WHO Consultation Accessed. 2005. Available online: http://apps.who.int/iris/bitstream/10665/44027/1/9789241597470_eng.pdf (accessed on 2 March 2020)

15. Akoto, O.; Andoh, H.; Darko, G.; Eshun, K.; Osei-Fosu, P. Health risk assessment of pesticides residue in maize and cowpea from ejura, Ghana. *Chemosphere* **2013**, *92*, 67–73. [CrossRef]

16. MOA. *GB 2763-2021*; National Food Safety Standard Maximum Residue Limits for Pesticides in Food. Standards Press of China: Beijing, China, 2021. (In Chinese)

17. Li, Z.X.; Zhang, Y.H.; Zhao, Q.Y.; Wang, C.Q.; Cui, Y.L.; Li, J.; Chen, A.H.; Liang, G.L.; Jiao, B.N. Occurrence, temporal variation, quality and safety assessment of pesticide residues on citrus fruits in China. *Chemosphere* **2020**, *258*, 127381. [CrossRef]

18. Li, Z.X.; Zhang, Y.H.; Zhao, Q.Y.; Cui, Y.L.; He, Y.; Li, J.; Yang Qin Lin, Z.H.; Wang, C.Q.; Liang, G.L.; Jiao, B.N. Determination, distribution and potential health risk assessment of insecticides and acaricides in citrus fruits of China. *J. Food Compos. Anal.* **2022**, *111*, 104645. [CrossRef]

19. Mac Loughlin, T.M.; Peluso, M.L.; Etchegoyen, M.A.; Alonso, L.L.; de Castro, M.C.; Percudani, M.C.; Marino, D.J.G. Pesticide residues in fruits and vegetables of the argentine domestic market: Occurrence and quality. *Food Control* **2018**, *93*, 129–138. [CrossRef]

20. WHO. Global Environment Monitoring System (GEMS)/Food Contamination Monitoring and Assessment Programme. 2021. Available online: https://www.who.int/teams/nutrition-and-food-safety/databases/global-environment-monitoring-system-food-contamination (accessed on 30 September 2021).

21. WHO. Inventory of Evaluations Performed by the Joint Meeting on Pesticide Residues (JMPR). 2012. Available online: http://apps.who.int/pesticideresidues-jmpr-database/Home/Range/All (accessed on 12 August 2021).

22. MOA. *GB/T 6379.1-2004*; Accuracy (Trueness and Precision) of Measurement Methods and Results-Prat 1: General Principles and Definitions. Standards Press of China: Beijing, China, 2004. (In Chinese)

23. MOA. *GB/T 6379.2-2004*; Accuracy (Trueness and Precision) of Measurement Methods and Results-Prat 2: Basic Method for the Determination of Repeatability and Reproducibility of a Standard Measure Method. Standards Press of China: Beijing, China, 2004. (In Chinese)

24. Hao, W.N.; Li, H.; Hu, M.Y.; Yang, L.; Rizwan-ul-Haq, M. Integrated control of citrus green and blue mold and sour rot by Bacillus amyloliquefaciens in combination with tea saponin. *Postharvest Biol. Technol.* **2011**, *59*, 316–323. [CrossRef]

25. Livingston, G.; Hack, L.; Steinmann, K.P.; Graftoncardwell, E.E.; Rosenheim, J.A. An ecoinformatics approach to field-scale evaluation of insecticide effects in California citrus: Are citrus thrips and citrus red mite induced pests? *J. Econ. Entomol.* **2018**, *111*, 1290–1297. [CrossRef]

26. Bakırcı, G.T.; Acay, D.B.Y.; Bakırcı, F.; Otles, S. Pesticide residues in fruits and vegetables from the Aegean region, Turkey. *Food Chem.* **2014**, *160*, 379–392. [CrossRef]

27. Li, Z.X.; Nie, J.Y.; Lu, Z.Q.; Xie, H.Z.; Kang, L.; Chen, Q.S.; Li, A.; Zhao, X.B.; Xu, G.F.; Yan, Z. Cumulative risk assessment of the exposure to pyrethroids through fruits consumption in China-Based on a 3-year investigation. *Food Chem. Toxicol.* **2016**, *96*, 234–243. [CrossRef]

28. Mutengwe, M.T.; Chidamba, L.; Korsten, L. Monitoring pesticide residues in fruits and vegetables at two of the biggest fresh produce markets in Africa. *J. Food Prot.* **2016**, *79*, 1938–1945. [CrossRef]

29. Li, M.M.; Dai, C.; Wang, F.Z.; Kong, Z.Q.; He, Y.; Huang, Y.T.; Fan, B. Chemometric-assisted QuEChERS extraction method for post-harvest pesticide determination in fruits and vegetables. *Sci. Rep.* **2017**, *7*, 42489. [CrossRef]

30. MOA. *GB 2763-2014*; National Food Safety Standard Maximum Residue Limits for Pesticides in Food. Standards Press of China: Beijing, China, 2014. (In Chinese)

31. Li, Z.X.; Nie, J.Y.; Yan, Z.; Xu, G.F.; Li, H.F.; Kuang, L.X.; Pan, L.G.; Xie, H.Z.; Wang, C.; Liu, C.D. Risk assessment and ranking of pesticide residues in Chinese pears. *J. Integr. Agric.* **2015**, *14*, 2328–2339. [CrossRef]

32. Suarez-Jacobo, A.; Alcantar-Rosales, V.M.; Alonso-Segura, D.; Heras-Ramírez, M.; Elizarragaz-De La Rosa, D.; Lugo-Melchor, O.; Gaspar-Ramirez, O. Pesticide residues in orange fruit from citrus orchards in Nuevo Leon State, Mexico. *Food Addit. Contam. Part B* **2017**, *10*, 192–199. [CrossRef]

33. EFSA. The 2017 European Union report on pesticide residues in food. *EFSA J.* **2019**, *17*, e05743.

34. Poulsen, M.E.; Andersen, J.H.; Petersen, A.; Jensen, B.H. Results from the Danish monitoring programme for pesticide residues from the period 2004–2011. *Food Control* **2017**, *74*, 25–33. [CrossRef]

35. MOA. *GB 2763-2020*; National Food Safety Standard Maximum Residue Limits for Pesticides in Food. Standards Press of China: Beijing, China, 2020. (In Chinese)

36. Jardim, A.N.O.; Mello, D.C.; Goes, F.C.S.; Junior, E.F.F.; Caldas, E.D. Pesticide residues in cashew apple, guava, kaki and peach: GC-mECD, GC-FPD and LC-MS/MS multiresidue method validation, analysis and cumulative acute risk assessment. *Food Chem.* **2014**, *164*, 195–204. [CrossRef]

37. Szpyrka, E.; Kurdziel, A.; Matyaszek, A.; Podbielska, M.; Rupar, J.; Słowik-Borowiec, M. Evaluation of pesticide residues in fruits and vegetables from the region of south-eastern Poland. *Food Control* **2015**, *48*, 137–142. [CrossRef]

38. Jensen, B.H.; Petersen, A.; Nielsen, E.; Christensen, T.; Poulsen, M.; Andersen, J.H. Cumulative dietary exposure of the population of Denmark to pesticides. *Food Chem. Toxicol.* **2015**, *83*, 300–307. [CrossRef]
39. Chung, S.W. How effective are common household preparations on removing pesticide residues from fruit and vegetables? A review. *J. Sci. Food Agric.* **2018**, *98*, 2857–2870. [CrossRef]
40. Jankowska, M.; Łozowicka, B.; Kaczynski, P. Comprehensive toxicological study over 160 processing factors of pesticides in selected fruit and vegetables after water, mechanical and thermal processing treatments and their application to human health risk assessment. *Sci. Total Environ.* **2019**, *652*, 1156–1167. [CrossRef]

Article

Spatial Distribution and Migration Characteristic of Forchlorfenuron in Oriental Melon Fruit by Matrix-Assisted Laser Desorption/Ionization Mass Spectrometry Imaging

Qi Wang [1], Xiaohui Li [1], Hongping Wang [1], Simeng Li [1], Chen Zhang [1], Xueying Chen [1], Jing Dong [2], Hua Shao [1], Jing Wang [1] and Fen Jin [1,*]

[1] Key Laboratory of Agro-Product Quality and Safety, Institute of Quality Standards & Testing Technology for Agro-Products, Chinese Academy of Agricultural Sciences, Beijing 100081, China; wangqi2021yw@163.com (Q.W.); nkshaohua@163.com (H.S.)

[2] Shimadzu China MS Center, Beijing 100020, China

* Correspondence: jinfenbj@163.com or jinfen@caas.cn; Tel.: +86-010-82106506; Fax: +86-010-82106500

Abstract: Forchlorfenuron is a widely used plant growth regulator to support the pollination and fruit set of oriental melons. It is critical to investigate the spatial distribution and migration characteristics of forchlorfenuron among fruit tissues to understand its metabolism and toxic effects on plants. However, the application of imaging mass spectrometry in pesticides remains challenging due to the usually extremely low residual concentration and the strong interference from plant tissues. In this study, a matrix-assisted laser desorption/ionization mass spectrometry imaging (MALDI-MSI) method was developed for the first time to obtain the dynamic images of forchlorfenuron in oriental melon. A quantitative assessment has also been performed for MALDI-MSI to characterize the time-dependent permeation and degradation sites of forchlorfenuron in oriental melon. The majority of forchlorfenuron was detected in the exocarp and mesocarp regions of oriental melon and decreased within two days after application. The degradation rate obtained by MALDI-MSI in this study was comparable to that obtained by HPLC-MS/MS, indicating that the methodology and quantification approach based on the MALDI-MSI was reliable and practicable for pesticide degradation study. These results provide an important scientific basis for the assessment of the potential risks and effects of forchlorfenuron on oriental melons.

Keywords: forchlorfenuron; mass spectrometry imaging; matrix-assisted laser desorption/ionization; spatial distribution; migration characteristic

Citation: Wang, Q.; Li, X.; Wang, H.; Li, S.; Zhang, C.; Chen, X.; Dong, J.; Shao, H.; Wang, J.; Jin, F. Spatial Distribution and Migration Characteristic of Forchlorfenuron in Oriental Melon Fruit by Matrix-Assisted Laser Desorption/Ionization Mass Spectrometry Imaging. *Foods* **2023**, *12*, 2858. https://doi.org/10.3390/foods12152858

Academic Editor: Roberto Romero-González

Received: 9 June 2023
Revised: 17 July 2023
Accepted: 24 July 2023
Published: 27 July 2023

1. Introduction

Forchlorfenuron, a synthetic plant growth regulator, has been used in many horticulture plants to support pollination and increase productivity during the past 30 years [1,2]. Recently, due to its detection in different agro-products and toxicity, forchlorfenuron has received greater scientific and regulatory scrutiny. Forchlorfenuron has been associated with severe hydrometra in the uterus and pathological changes in the ovaries of Sprague–Dawley (SD) rats [3] and classified as a category 2 carcinogenic agent by the European Food Safety Authority [4–6]. Adverse effects on fruit quality have also been reported after forchlorfenuron application [7,8].

Investigating the spatial distribution of forchlorfenuron among fruit tissues is important for understanding its metabolism and toxic effects on plants. Some studies have been conducted on kiwifruit, grape, and apple, and the ^{14}C-forchlorfenuron exhibited different migration characteristics in a variety of fruits [1]. However, the method lacks molecular specificity and relies on radiolabeled tracers [9–11]. In recent years, high-performance liquid chromatography–mass spectrometry (HPLC-MS) has also been reported as an alternative method to study the spatial distribution of forchlorfenuron [12], whereas the necessary tissue homogenates procedure for determination can still not to achieve the in situ analysis of forchlorfenuron in the sample tissue. To better understand the spatial distribution and migration characteristic of forchlorfenuron in fruit, visual information is more useful than chromatographic data. It provides an anatomical distribution of forchlorfenuron inside the fruit.

Mass spectrometry imaging (MSI) coupled with matrix-assisted laser desorption/ionization (MALDI) has proven to be a promising technique for spatial analysis with its high molecular specificity and visualization for endogenous metabolites and exogenous medicines [13–15]. However, analyzing pesticides in plant tissues is not easy due to the trace residual concentration and the strong interference from the MALDI matrix [16,17]. It is well-known that in MALDI-MSI, the matrix and its coating pattern can have a significant influence on ionization efficiency in the MALDI ion source, and an internal standard can control for the irreproducibility of ion signals from scan to scan for small-molecule compounds. Therefore, the matrix selection and its coating pattern are the critical steps to obtain high-quality MALDI mass spectrometry (MS) signals for several small-molecule compounds, including sulfonylurea herbicides and fungicide metalaxyl in recent years [18–21].

Oriental melon (*Cucumis melo var. makuwa*) is an important fruit worldwide, and more than 28 million tons were produced in 2020 (FAOSTAT 2020). In China, more than 42.1% of oriental melons are treated with forchlorfenuron to promote fruit set and increase fruit weight [22]. To the best of our knowledge, no investigation of the spatial distribution of forchlorfenuron in oriental melons has been reported. In this study, we developed an MALDI-MSI method to study the spatial distribution and time-dependent migration of forchlorfenuron in oriental melon fruit. Considering the effective factors on the sensitivity, the matrix and matrix coating pattern were optimized. The isotope internal standard was applied to improve the quantitative capabilities of MSI. This is the first report about the dynamic images of forchlorfenuron in oriental melon and we hope that it will direct future studies on the fate of pesticides in fruit tissues.

2. Material and Methods

2.1. Chemicals and Reagents

The standard of forchlorfenuron (purity > 98%) was purchased from Dr. Ehrenstorfer (Augsburg, Germany). Forchlorfenuron-d5, the isotopic internal standard (IS), was purchased from Toronto Research Chemicals (Toronto, ON, Canada). HPLC-grade methanol, formic acid, and pure water were obtained from Fisher (Marshalltown, IA, USA). MALDI-grade α-cyano-4-hydroxycinnamic acid (CHCA), 2,5-dihydroxybenzoic acid (DHB), and 9-aminoacridine (9-AA) were purchased from Sigma-Aldrich (St. Louis, MO, USA). Other reagents and solvents used in the present study were of analytical grade.

2.2. Preparation of Standard Solutions and Matrix Solutions

The stock solution of 100 mg/L forchlorfenuron standard was prepared in MeOH and diluted to the needed concentrations before use. For the matrix deposition on the sample sections, DHB and 9-AA were prepared in 10 mg/mL in MeOH/water (70/30, *v/v*). In total, 50 mg/mL of CHCA was prepared in MeOH/water (70/30, *v/v*).

2.3. Field Trials and Sample Collection

The field trial was conducted on oriental melons (*Cucumis melo var. makuwa*) from July to October 2018 in a greenhouse at the Chinese Academy of Agricultural Sciences. The field trials were divided into a treatment group and a control group. The dose of forchlorfenuron soluble concentrate (SL) was set to 20 mg/L according to the recommended dose on the registered label. In the treatment group, 20 mg/L forchlorfenuron solution was used to dip the flower and fruit ovary for 1–2 s during the flowering stage with BBCH code 61 of the oriental melon. A separate plot for the no-forchlorfenuron application was used as a control, and oriental melons required manual pollination to generate homozygous parental lines. At least 6 representative oriental melon fruit samples were collected randomly from each plot at 2 h, 1 d, 3 d, and 4 d after pollination or application of forchlorfenuron. All samples were flash-frozen in liquid nitrogen and stored at $-80\ ^\circ$C until analysis.

2.4. Tissue Sectioning and Matrix Deposition for MALDI TOF-MSI

Frozen oriental melon samples were cut longitudinally into 35 μm sections and thaw-mounted on indium tin oxide glass slides for MSI at $-18\ ^\circ$C using a cryostat microtome (Leica CM1950, Nussloch, Germany). Matrix coating modes were optimized by comparing the signal/noise (S/N) of forchlorfenuron produced under 3 possible matrix coating procedures as follows: (1) Sublimation: 300 mg CHCA was applied by vacuum sublimation using the iMLayer system (Sanyu Electron, Tokyo, Japan) for 8 min at 250 $^\circ$C. (2) Airbrushing: 500 μL CHCA matrix solution (10 mg/mL, in MeOH and distilled water (v/v, 7:3) was sprayed onto the tissue surface by airbrush. (MR. Linear Compressor L7/PS270 Airbrush, Tokyo, Japan). The distance between the airbrush tip and the tissue surface was about 8 cm. For the first 3 cycles, the matrix was airbrushed for 2 s at 60 s intervals and, in the following 7 cycles, the matrix was continuously sprayed for 1 s at 30 s intervals. After spraying, the glass slide was air-dried for 5 min to vaporize the solvent. (3) A "two step matrix application" method was used: sublimation combined with an airbrushing step. The first step was the same as that described in the above "(1) Sublimation" part. For the second step, 1 mL matrix (10 mg/mL) solution or methanol was sprayed onto the glass slide for 10 cycles using the airbrush according to the procedures described in "(2) Airbrushing". Finally, the sample sections were air-dried for 2 min to vaporize the solvent.

2.5. Mass Spectrometry Imaging Conditions

Mass spectrometry imaging was performed using a mass microscope (iMScope TRIO, Shimadzu, kyoto, Japan) equipped with a high-resolution optical microscope and a hybrid ion trap time-of-flight (IT-TOF) mass spectrometer with an atmospheric pressure MALDI with a diode-pumped 355 nm Nd: YAG laser (Shimadzu Corporation, Kyoto, Japan). Optical images and ion distribution data under the same microscope were obtained. The laser power was 45 eV and 55 eV for the precursor ion (m/z 248.05) and the product ions (m/z 129.02, m/z 155.00) of forchlorfenuron, respectively. The mass spectrometry conditions for the precursor ion (MS stage 1) and the product ion (MS stage 2) of forchlorfenuron were as follows:

MS stage 1: sample voltage, 3.5 kV; detector voltage, 1.95 kV; number of laser shots, 80; repetition rate, 1000 Hz; laser diameter, 2 (20 μm); ion polarity, positive; mass range, 200–300; laser intensity, 45 eV.

MS stage 2: sample voltage, 3.5 kV; detector voltage, 1.95 kV; number of laser shots, 80; repetition rate, 1000 Hz; laser diameter, 2 (20 μm); precursor ion, m/z 248.056; mass range, 50–300; laser intensity, 55 eV.

2.6. Quantitative Analysis

Standard curve: A series of forchlorfenuron standard solutions (0.1, 0.5, 1.0, 5.0, 10.0, 20.0 mg/L) containing 5 mg/L forchlorfenuron-d5 isotopic IS solution were deposited onto the blank oriental melon sections before matrix application using a micropipettor (1 μL); each concentration was repeated three times. The standard curve equation was plotted using the ion intensity extracted from the regions of the standard point of Qual Browser to Microsoft Excel.

Sample quantification: Methods of internal standard application were adapted from Chumbley et al. [23] All oriental melon tissue sections received IS (5 mg/L forchlorfenuron-d5 isotopic IS solution, 1μL) in different regions (n = 9) before the sublimation of the matrix. The extracted ion currents for forchlorfenuron (m/z 248.05→129.02) and forchlorfenuron-d5 (m/z 253.05→129.02) were plotted using data extracted from the regions of interest (ROI) of oriental melon sections to Microsoft Excel.

2.7. Statistical Analysis

The data acquisition, visualization, and quantification were performed by the imaging MS Solution Version 1.30 Software (Shimadzu Corporation, Tokyo, Japan). Regions of interest (ROI) were manually defined in the analysis software using the optical and MSI data image. The data are expressed as the mean value of three replicates with the standard deviation (SD).

3. Results and Discussion

3.1. Optimization of Matrix

Based on the fact that the matrix is one of the most important characteristics influencing sensitivity in MALDI-MSI, the effects of three matrices (DHB, CHCA, and 9-AA) on the ionization and sensitivity of forchlorfenuron were investigated. As shown in Figure 1, CHCA produced the highest signal responses of the target precursor ion (m/z 248.05) and product ions (m/z 129.02, m/z 155.00) of forchlorfenuron, which were 8–15 times higher than those obtained by using 9AA or DHB.

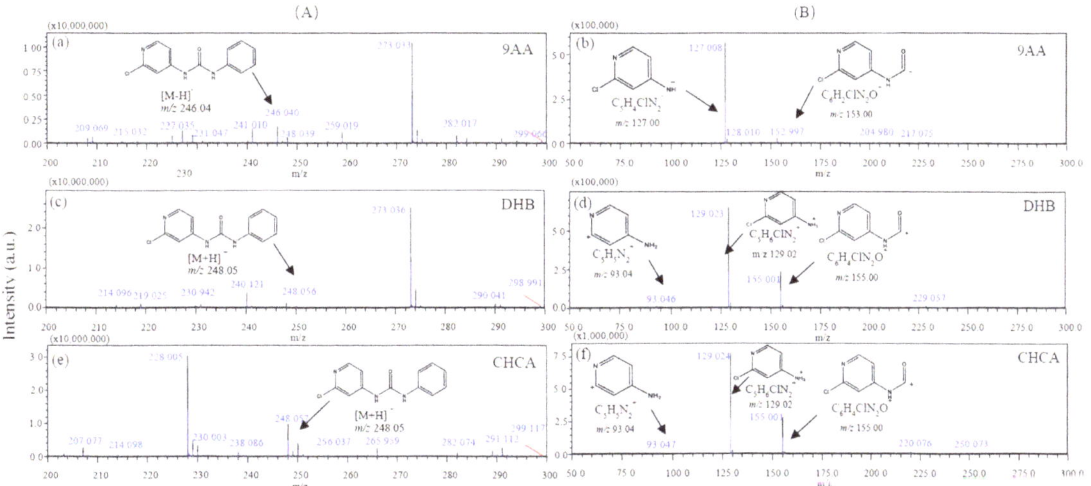

Figure 1. The structures and mass spectra of forchlorfenuron by MALDI-MSI with different matrices, the forchlorfenuron standard solution (1 µL, 100 mg/L) deposited on blank oriental melon sections (2 h after manual pollination) before matrix sublimation. (**A**) Precursor ion mass spectrum. (**B**) Product ion mass spectrum. (**a,b**) 9AA in the negative mode. (**c,d**) DHB in the positive mode. (**e,f**) CHCA in the positive mode.

In addition, the background interference of these matrices in MSI analysis was evaluated by sublimating these three matrices on the oriental melon sections from the control group and treatment group (2 h after manual pollination or application of the forchlorfenuron), respectively. As shown in Figure 2I, the interfering signals were obviously observed at all precursor ions (m/z 248.05, Figure 2Ia–c) and product ions (m/z 155.00, Figure 2Ig) using the DHB matrix. However, compared with control groups (Figure 2I), the relative intensity and profiles of the signals of forchlorfenuron were significantly increased and clearly observed in treatment oriental melon sections (Figure 2II). Moreover, the CHCA matrix showed a higher signal intensity, provided the clear visualization of forchlorfenuron, and avoided background interferences from matrix and melon tissues in the images for product ions (m/z 129.02 and m/z 155.00). These results suggested the high selectivity of the CHCA matrix for forchlorfenuron detection in tissue sections by MALDI-MSI. Furthermore, with less background and high sensitivity, the secondary product ions (m/z 129.02 and m/z 155.00) of forchlorfenuron were selected for subsequent analysis.

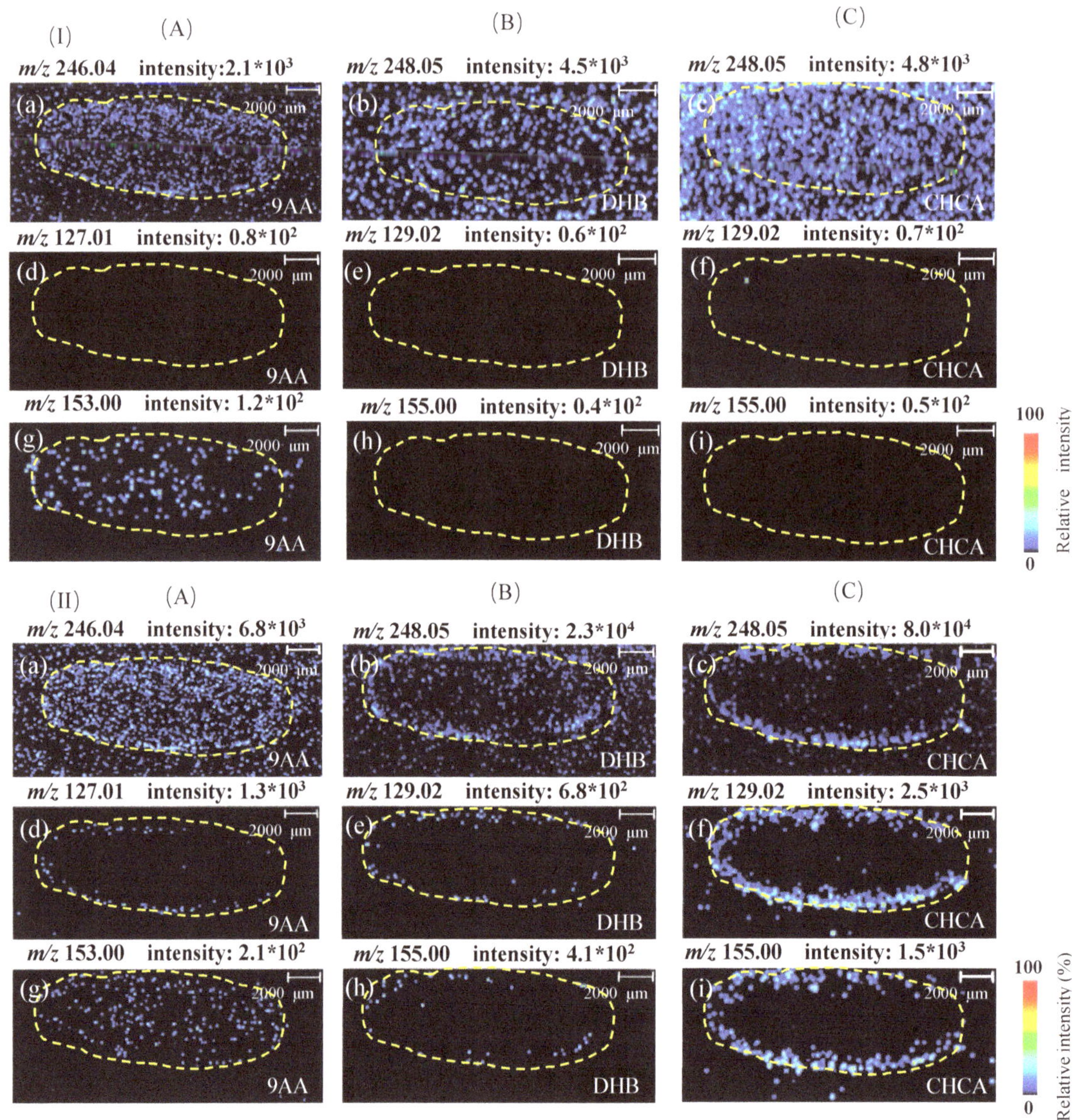

Figure 2. Representative ion images of forchlorfenuron in oriental melon sections in control group (**I**) and treatment group (**II**) by MAIDL-MSI with different matrices. (**A**) 9AA in the positive mode. (**a**) precursor ion. (**d**) product ion 1. (**g**) product ion 2. (**B**) DHB in positive ion mode. (**b**) Precursor ion. (**e**) product ion 1. (**h**) product ion 2. (**C**) CHCA in positive ion mode. (**c**) Precursor ion. (**f**) product ion 1. (**i**) product ion 2.

3.2. Optimization of Matrix Application Methods

The matrix preparation and deposition procedure are the other critical parts of a successful MALDI-IMS analysis [24,25]. In this study, in order to obtain the maximum ion signal intensity and good-quality images of forchlorfenuron, different matrix coating modes such as matrix sublimation and manual spraying were adopted for the pretreatment of the tissue sections containing 100 mg/L of forchlorfenuron. As shown in Figure 3, compared to the manual spraying of CHCA, the sublimation CHCA method provided higher signal intensities of forchlorfenuron, suggesting that this method produced relatively small crystallization and homogeneous matrix–forchlorfenuron cocrystal, which was similar to the results reported in the previous paper [14,16,18,26]. In addition, an attempt to further improve the sensitivity of forchlorfenuron was made by the "two-step matrix application", which combined matrix sublimation and the airbrushing method used in this study. When CHCA was used in the second step by airbrushing, it was found that there was a decrease in the precursor ion and the two product ions in comparison to the one-step method, which was possibly associated with the uneven distribution and larger CHCA matrix crystal on the oriental melon tissue. However, when methanol was used in the second step by airbrushing, as expected, it produced a better sensitivity than that used in CHCA. It was possibly related to the increase in the rate of matrix–analyte co-crystallization by spraying methanol, which was similar to the results reported in an analysis of octreotide in the mouse target tissue [27]. Therefore, the two-step matrix coating method, combining the sublimation of CHCA and manual spraying of methanol, was adopted in the subsequent study.

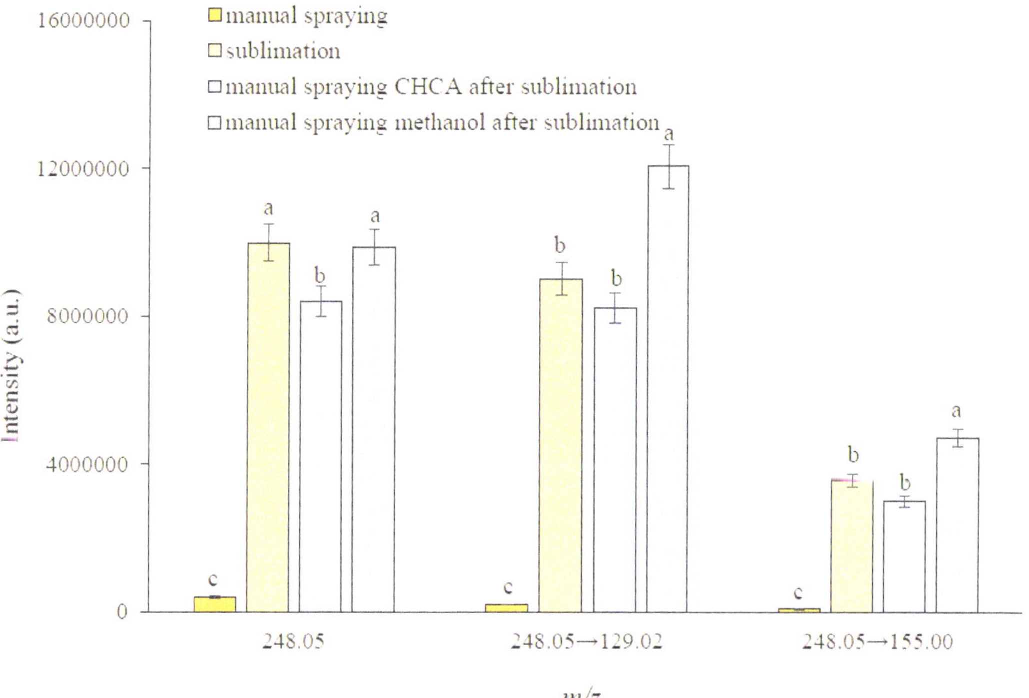

Figure 3. Comparison of detection sensitivities of forchlorfenuron after different matrix coating modes; forchlorfenuron standard solution (1 µL, 100 mg/L) was deposited on the blank oriental melon section (n = 3). Bars that do not share similar letters denote statistically significant differences ($p < 0.05$) determined by a pairwise t-test with Bonferroni-adjusted p-values.

3.3. Method Validation

To validate the developed MSI analytical method, the spatial distribution of forchlorfenuron was examined in oriental melon fruit from the control group (2 d after chasmogamy) and treatment group (2 d after application of forchlorfenuron) in field trials. In the control group (Figure 4A), forchlorfenuron was scarcely detected, as illustrated by the imaging of both secondary product ions. In the application group (Figure 4B), the profiles of the signals of product ions (m/z 129.02 and m/z 155.00) of forchlorfenuron were clearly distinguished from the background noise, and the intensities of the product ions (m/z 129.02 and m/z 155.00) of forchlorfenuron were 2.5×10^3 and 1.5×10^3, respectively. This result indicated that the optimized method could be used to investigate the spatial distribution of forchlorfenuron in melon fruits.

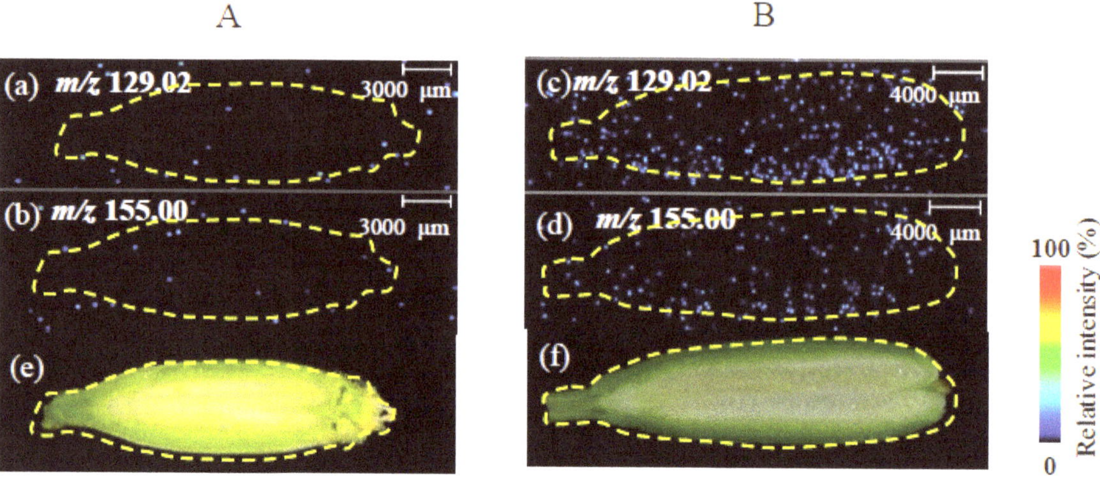

Figure 4. Distribution of forchlorfenuron in oriental melon, as shown by iMScope, in (**A**) control group (2 d after chasmogamy) and (**B**) treatment group (2 d after application of forchlorfenuron). (**a–d**) Product ion mass imaging MS of m/z 129.02, 155.00. (**e**,**f**) Optical image of oriental melon slides.

3.4. Quantitative Determination of Forchlorfenuron in Oriental Melon by MSI

Figure 5A showed a calibration curve by plotting the average intensity of forchlorfenuron against the concentrations on oriental melon sections. There was a positive linear correlation between signal intensity and the concentration of forchlorfenuron between 0.10 mg/L and 20.0 mg/L ($n = 3$), with a linear correlation coefficient (R^2) of 0.8198. The intensity ratio of forchlorfenuron/forchlorfenuron-d5 against the concentrations on oriental melon sections was shown in Figure 5B. The linear correlation coefficient (R^2) was expected to improve to 0.9945 by normalizing the forchlorfenuron/forchlorfenuron-d5 ion signal ratio for quantitative analysis, and the relative standard deviations of the measurements from each spot in the tissue section were less than 13.0%. Linearity and scan-to-scan reproducibility were significantly improved when normalizing by the IS. These improvements can be attributed to the control for ionization variability from the matrix and tissue heterogeneity, inefficient analyte extraction, and ionization suppression effects in tissue surfaces [28].

Furthermore, as shown in Figure 5C, the imaging bright changes in spotted forchlorfenuron solutions also matched well with the forchlorfenuron standard concentrations from 0.10 mg/L to 20.0 mg/L and maintained stability in the d5-forchlorfenuron solutions of constant concentration (Figure 5D). Therefore, MALDI-MSI was shown to be suitable for the quantitative analysis of forchlorfenuron in oriental melon after internal calibration. Multiple analyses of replicate tissue sections from multiple melons produced reproducible quantitative results. Therefore, MSI was shown to be suitable for the quantitative analysis of forchlorfenuron in oriental melon, and the LOQ was found to be 0.1 mg/kg.

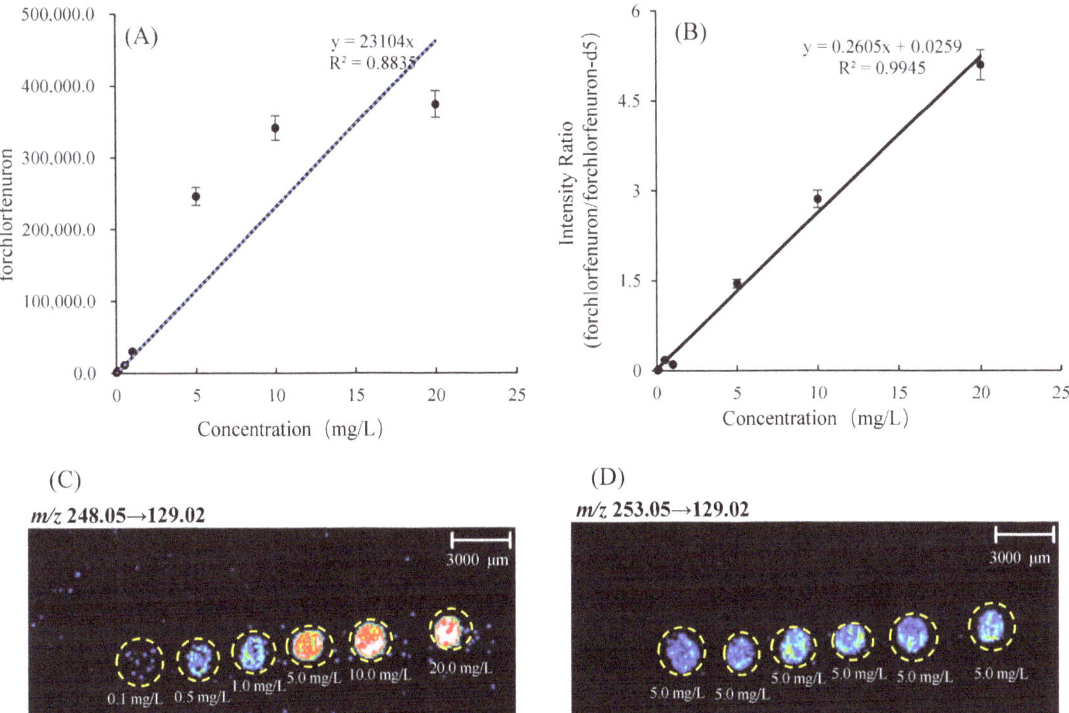

Figure 5. MALDI-MSI experiment for quantitative analysis of forchlorfenuron from oriental melon tissue. Calibration curve (n = 3) of forchlorfenuron generated using (**A**) the average intensity of m/z 129.02 and (**B**) the ratio average intensity of m/z 248.05→129.02/253.05→129.02. (**C**) Ion intensity of m/z 248.05→129.02 of forchlorfenuron was used to generate MS/MS images. Calibration spots increase in concentration from 0.10 to 20.0 mg/L. (**D**) Ion intensity of m/z 253.05→129.02 of d5-forchlorfenuron (5.0 mg/L) was used to generate MS/MS images.

3.5. Spatial Distribution and Migration Characteristics of Forchlorfenuron in Oriental Melon by MALDI-MSI

The developed MALDI-MSI method was used to investigate the spatial distribution and time-dependent permeation of forchlorfenuron in oriental melon fruit in field trails. Figure 6A shows the optical image of a longitudinal cross section of an oriental melon fruit under the microscope of the iMScope instrument.

As shown in Figure 6B, a gradual and continuous migration and degradation of forchlorfenuron in oriental melon occurred between 2 h and 4 days after application. It was found that forchlorfenuron was mainly distributed in the exocarp region of oriental melon fruit, and the concentration was estimated at 0.8 mg/kg after 2 h. After one day, forchlorfenuron generally migrated from the exocarp into the mesocarp region with a concentration of 0.4 mg/kg. After two days, most forchlorfenuron gradually migrated from the mesocarp into the endocarp region with a concentration of 0.1 mg/kg, indicating that forchlorfenuron degraded quickly and penetrated in oriental melon. After four days, no distinct signal was observed in the imaging of the forchlorfenuron ions observed, which means that the concentrations of forchlorfenuron were lower than 0.1 mg/kg. It is maybe related to the metabolism of 4-hydroxyphenyl-forchlorfenuron in oriental melon fruit [29]. The degradation rate obtained by MALDI-MSI in this study was comparable to that obtained by HPLC-MS/MS in our previous study [29]. In our previous paper, the half-life of forchlorfenuron in oriental melon fruit was 1.29 days, and 4-hydroxyphenyl-forchlorfenuron was first detected at 4d with concentrations of 4.5 μg/kg by HPLC-MS/MS.

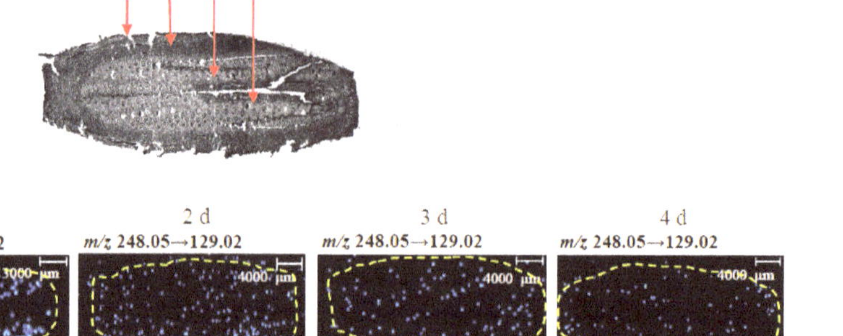

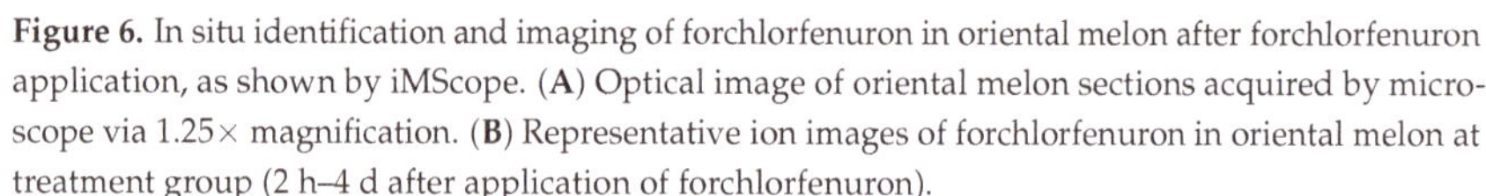

Figure 6. In situ identification and imaging of forchlorfenuron in oriental melon after forchlorfenuron application, as shown by iMScope. (**A**) Optical image of oriental melon sections acquired by microscope via 1.25× magnification. (**B**) Representative ion images of forchlorfenuron in oriental melon at treatment group (2 h–4 d after application of forchlorfenuron).

The radioactive residues of ^{14}C-forchlorfenuron were major in the skin fractions of kiwifruit after 127 days, while approximately 62% of the radioactive residues were associated with the apple pulp and skin after 114 days. In our study, the majority of forchlorfenuron was detected in the exocarp and mesocarp regions of oriental melon and decreased within two days after application. These results may be related to the increasing flesh thickness of oriental melon post-harvest after forchlorfenuron treatment [30]. These results can help us better understand the effect and biological fate of forchlorfenuron in oriental melon.

4. Conclusions

In conclusion, an MALDI-MSI method was developed for the first time to obtain the dynamic images of forchlorfenuron in oriental melon. The matrix and matrix coating methods were optimized to improve the sensitivity of forchlorfenuron in melon tissue greatly. Good quantitative accuracy and sharp images were obtained when plotting characteristic fragment ions normalized with IS. The method was successfully applied to investigate the spatial distribution and time-dependent permeation of forchlorfenuron in oriental melon fruit after forchlorfenuron treatment. Most forchlorfenuron was detected in the exocarp and mesocarp regions within two days after application, and the overall concentration gradually decreased over time. The degradation rate obtained by MALDI-MSI in this study was comparable to that obtained by HPLC-MS/MS, which provides an important scientific basis for assessing the potential risks to consumers and effects of forchlorfenuron on oriental melons.

Author Contributions: Q.W.: conceptualization, methodology, validation, formal analysis. X.L. and H.W.: methodology, validation. S.L. and C.Z.: visualization. J.D. and X.C.: investigation. H.S.: visualization, investigation. J.W.: supervision. F.J.: conceptualization, supervision, project administration, funding acquisition. All authors have read and agreed to the published version of the manuscript.

Funding: This work was supported by the National Natural Science Foundation of China (No. 31871890) and the Central Public-interest Scientific Institution Basal Research Fund (No. Y2022XK29).

Data Availability Statement: The data used to support the findings of this study can be made available by the corresponding author upon request.

Acknowledgments: The authors gratefully acknowledge the Agricultural Science and Technology Innovation Program and the Leading Talent Program under the Chinese Academy of Agricultural Science.

Conflicts of Interest: Author Jing Dong was employed by the company Shimadzu China. The authors gratefully acknowledge Shimadzu China company for providing the materials and equipment. The remaining authors declare that the research was conducted in the absence of any commercial or financial relationships that could be construed as a potential conflict of interest.

References

1. Pesticides, A.; Authority, V.M. *Public Release Summary on Evaluation of the New Active Forchlorfenuron in the Product Sitofex 10EC Plant Growth Regulator*; Australian Pesticides and Veterinary Medicines Authority: Symonston, Australia, 2005.
2. Shan, T.T.; Zhang, X.; Guo, C.F.; Guo, S.H.; Zhao, X.B.; Yuan, Y.H.; Yue, T.L. Identity, Synthesis, and Cytotoxicity of Forchlorfenuron Metabolites in Kiwifruit. *J. Agric. Food Chem.* **2021**, *69*, 9529–9535. [CrossRef] [PubMed]
3. Bu, Q.; Wang, X.Y.; Xie, H.C.; Zhong, K.; Wu, Y.P.; Zhang, J.Q.; Wang, Z.S.; Gao, H.; Huang, Y.N. 180 Day Repeated-Dose Toxicity Study on Forchlorfenuron in Sprague-Dawley Rats and Its Effects on the Production of Steroid Hormones. *J. Agric. Food Chem.* **2019**, *67*, 10207–10213. [CrossRef] [PubMed]
4. Zhang, Z.W.; Guo, K.Q.; Bai, Y.B.; Dong, J.; Gao, Z.H.; Yuan, Y.H.; Wang, Y.; Liu, L.P.; Yue, T.L. Identification, Synthesis, and Safety Assessment of Forchlorfenuron (1-(2-Chloro-4-pyridyl)-3-phenylurea) and Its Metabolites in Kiwifruits. *J. Agric. Food Chem.* **2015**, *63*, 3059–3066. [CrossRef]
5. Gong, G.; Kam, H.; Tse, Y.; Lee, S.M. Cardiotoxicity of forchlorfenuron (CPPU) in zebrafish (Danio rerio) and H9c2 cardiomyocytes. *Chemosphere* **2019**, *235*, 153–162. [CrossRef]
6. European Food Safety Authority (EFSA); Arena, M.; Auteri, D.; Barmaz, S.; Bellisai, G.; Brancato, A.; Brocca, D.; Bura, L.; Byers, H.; Chiusolo, A.; et al. Peer review of the pesticide risk assessment of the active substance forchlorfenuron. *Efsa J.* **2017**, *15*, e04874. [CrossRef]
7. Qian, C.L.; Ren, N.N.; Wang, J.Y.; Xu, Q.; Chen, X.H.; Qi, X.H. Effects of exogenous application of CPPU, NAA and GA(4+7) on parthenocarpy and fruit quality in cucumber (Cucumis sativus L.). *Food Chem.* **2018**, *243*, 410–413. [CrossRef]
8. Ainalidou, A.; Karamanoli, K.; Menkissoglu-Spiroudi, U.; Diamantidis, G.; Matsi, T. CPPU treatment and pollination: Their combined effect on kiwifruit growth and quality. *Sci. Hortic.* **2015**, *193*, 147–154. [CrossRef]
9. Giepmans, B.N.; Adams, S.R.; Ellisman, M.H.; Tsien, R.Y. The fluorescent toolbox for assessing protein location and function. *Science* **2006**, *312*, 217–224. [CrossRef] [PubMed]
10. Glunde, K.; Jacobs, M.A.; Pathak, A.P.; Artemov, D.; Bhujwalla, Z.M. Molecular and functional imaging of breast cancer. *NMR Biomed.* **2009**, *22*, 92–103. [CrossRef]
11. Zierhut, M.L.; Ozturk-Isik, E.; Chen, A.P.; Park, I.; Vigneron, D.B.; Nelson, S.J. H-1 Spectroscopic Imaging of Human Brain at 3 Tesla: Comparison of Fast Three-Dimensional Magnetic Resonance Spectroscopic Imaging Techniques. *J. Magn. Reson. Imaging* **2009**, *30*, 473–480. [CrossRef]
12. Chen, W.J.; Jiao, B.N.; Su, X.S.; Zhao, Q.Y.; Qin, D.M.; Wang, C.Q. Dissipation and Residue of Forchlorfenuron in Citrus Fruits. *B Environ. Contam. Tox.* **2013**, *90*, 756–760. [CrossRef]
13. Matros, A.; Mock, H.-P. Mass Spectrometry Based Imaging Techniques for Spatially Resolved Analysis of Molecules. *Front. Plant Sci.* **2013**, *4*, 89. [CrossRef]
14. Mullen, A.K.; Clench, M.R.; Crosland, S.; Sharples, K.R. Determination of agrochemical compounds in soya plants by imaging matrix-assisted laser desorption/ionisation mass spectrometry. *Rapid Commun. Mass Spectrom.* **2005**, *19*, 2507–2516. [CrossRef]
15. Yoshimura, Y.; Zaima, N.; Moriyama, T.; Kawamura, Y. Different Localization Patterns of Anthocyanin Species in the Pericarp of Black Rice Revealed by Imaging Mass Spectrometry. *PLoS ONE* **2012**, *7*, 285. [CrossRef]
16. Enomoto, H.; Sato, K.; Miyamoto, K.; Ohtsuka, A.; Yamane, H. Distribution Analysis of Anthocyanins, Sugars, and Organic Acids in Strawberry Fruits Using Matrix-Assisted Laser Desorption/Ionization-Imaging Mass Spectrometry. *J. Agric. Food Chem.* **2018**, *66*, 4958–4965. [CrossRef] [PubMed]
17. van Kampen, J.J.A.; Burgers, P.C.; de Groot, R.; Gruters, R.A.; Luider, T.M. Biomedical Application of Maldi Mass Spectrometry for Small-Molecule Analysis. *Mass Spectrom. Rev.* **2011**, *30*, 101–120. [CrossRef] [PubMed]

18. Rao, T.; Shao, Y.H.; Hamada, N.; Li, Y.M.; Ye, H.; Kang, D.; Shen, B.Y.; Li, X.N.; Yin, X.X.; Zhu, Z.P.; et al. Pharmacokinetic study based on a matrix-assisted laser desorption/ionization quadrupole ion trap time-of-flight imaging mass microscope combined with a novel relative exposure approach: A case of octreotide in mouse target tissues. *Anal. Chim. Acta* **2017**, *952*, 71–80. [CrossRef] [PubMed]

19. Anderson, D.M.G.; Carolan, V.A.; Crosland, S.; Sharples, K.R.; Clench, M.R. Examination of the translocation of sulfonylurea herbicides in sunflower plants by matrix-assisted laser desorption/ionisation mass spectrometry imaging. *Rapid Commun. Mass Spectrom.* **2010**, *24*, 3309–3319. [CrossRef]

20. Pirman, D.A.; Yost, R.A. Quantitative Tandem Mass Spectrometric Imaging of Endogenous Acetyl-L-carnitine from Piglet Brain Tissue Using an Internal Standard. *Anal Chem.* **2011**, *83*, 8575–8581. [CrossRef]

21. Kubicki, M.; Lamshoft, M.; Lagojda, A.; Spiteller, M. Metabolism and spatial distribution of metalaxyl in tomato plants grown under hydroponic conditions. *Chemosphere* **2019**, *218*, 36–41. [CrossRef]

22. Roussos, P.A.; Ntanos, E.; Denaxa, N.K.; Tsafouros, A.; Bouali, I.; Nikolakakos, V.; Assimakopoulou, A. Auxin (triclopyr) and cytokinin (forchlorfenuron) differentially affect fruit physiological, organoleptic and phytochemical properties of two apricot cultivars. *Acta Physiol. Plant.* **2021**, *43*, 25. [CrossRef]

23. Chumbley, C.W.; Reyzer, M.L.; Allen, J.L.; Marriner, G.A.; Via, L.E.; Barry, C.E.; Caprioli, R.M. Absolute Quantitative MALDI Imaging Mass Spectrometry: A Case of Rifampicin in Liver Tissues. *Anal. Chem.* **2016**, *88*, 2392–2398. [CrossRef]

24. Han, J.; Permentier, H.; Bischoff, R.; Groothuis, G.; Casini, A.; Horvatovich, P. Imaging of protein distribution in tissues using mass spectrometry: An interdisciplinary challenge. *Trends Anal. Chem.* **2019**, *112*, 13–28. [CrossRef]

25. Shimma, S.; Takashima, Y.; Hashimoto, J.; Yonemori, K.; Tamura, K.; Hamada, A. Alternative two-step matrix application method for imaging mass spectrometry to avoid tissue shrinkage and improve ionization efficiency. *J. Mass Spectrom.* **2013**, *48*, 1285–1290. [CrossRef] [PubMed]

26. Gemperline, E.; Rawson, S.; Li, L. Optimization and Comparison of Multiple MALDI Matrix Application Methods for Small Molecule Mass Spectrometric Imaging. *Anal. Chem.* **2014**, *86*, 10030–10035. [CrossRef] [PubMed]

27. Tholey, A.; Heinzle, E. Ionic (liquid) matrices for matrix-assisted laser desorption/ionization mass spectrometry-applications and perspectives. *Anal. Bioanal. Chem.* **2006**, *386*, 24–37. [CrossRef] [PubMed]

28. Russo, C.; Brickelbank, N.; Duckett, C.; Mellor, S.; Rumbelow, S.; Clench, M.R. Quantitative Investigation of Terbinafine Hydrochloride Absorption into a Living Skin Equivalent Model by MALDI-MSI. *Anal. Chem.* **2018**, *90*, 10031–10038. [CrossRef]

29. Wang, Q.; Su, H.; Yue, N.; Li, M.J.; Li, C.M.; Wang, J.; Jin, F. Dissipation and risk assessment of forchlorfenuron and its major metabolites in oriental melon under greenhouse cultivation. *Ecotoxicol. Environ. Saf.* **2021**, *225*, 700. [CrossRef]

30. Jiang, Y.-L.; Chen, L.-Y.; Lee, T.-C.; Chang, P.-T. Improving postharvest storage of fresh red-fleshed pitaya (Hylocereus polyrhizus sp.) fruit by pre-harvest application of CPPU. *Sci. Hortic.* **2020**, *273*, 109646. [CrossRef]

Article

Development of Magnetic Lateral Flow and Direct Competitive Immunoassays for Sensitive and Specific Detection of Halosulfuron-Methyl Using a Novel Hapten and Monoclonal Antibody

Ying Ying [1,2,†], Xueyan Cui [1,2,†], Hui Li [1,2], Lingyi Pan [1,2], Ting Luo [1,2], Zhen Cao [1,2,*] and Jing Wang [1,2]

[1] Institute of Quality Standards and Testing Technology for Agro-Products, Key Laboratory of Agro-Product Quality and Safety, Chinese Academy of Agricultural Sciences, Ministry of Agriculture, Beijing 100081, China; yingying@caas.cn (Y.Y.); cxy19910411@163.com (X.C.); urhaan@163.com (L.P.); luoting074@163.com (T.L.); w_jing2001@126.com (J.W.)

[2] Institute of Quality Standards and Testing Technology for Agro-Products, Chinese Academy of Agricultural Sciences, Beijing 100081, China

* Correspondence: caozhen01@caas.cn

† These authors contributed equally to this work.

Abstract: Halosulfuron-methyl (HM) is widely used for the removal of noxious weeds in corn, sugarcane, wheat, rice, and tomato fields. Despite its high efficiency and low toxicity, drift to nontarget crops and leaching of its metabolites to groundwater pose potential risks. Considering the instability of HM, the pyrazole sulfonamide of HM was used to generate a hapten and antigen to raise a high-quality monoclonal antibody (Mab, designated 1A91H11) against HM. A direct competitive immunoassay (dcELISA) using Mab 1A91H11 achieved a half-maximal inhibitory concentration (IC50) of 1.5×10^{-3} mg/kg and a linear range of 0.7×10^{-3} mg/kg–10.7×10^{-3} mg/kg, which was 10 times more sensitive than a comparable indirect competitive ELISA (icELISA) and more simple to operate. A spiking recovery experiment performed in tomato and maize matrices with 0.01, 0.05, and 0.1 mg/kg HM had average recoveries within 78.9–87.9% and 103.0–107.4% and coefficients of variation from 1.1–6.8% and 2.7–6.4% in tomato and maize, respectively. In addition, a magnetic lateral flow immunoassay (MLFIA) was developed for quantitative detection of low concentrations of HM in paddy water. Compared with dcELISA, the MLFIA exhibited 3.3- to 50-fold higher sensitivity (IC50 0.21×10^{-3} mg/kg). The average recovery and RSD of the developed MLFIA ranged from 81.5 to 92.5% and 5.4 to 9.7%. The results of this study demonstrated that the developed dcELISA and MLFIA are suitable for rapid detection of HM residues in tomato and maize matrices and paddy water, respectively, with acceptable accuracy and precision.

Keywords: halosulfuron-methyl; monoclonal antibody; magnetic lateral flow immunoassay

Citation: Ying, Y.; Cui, X.; Li, H.; Pan, L.; Luo, T.; Cao, Z.; Wang, J. Development of Magnetic Lateral Flow and Direct Competitive Immunoassays for Sensitive and Specific Detection of Halosulfuron-Methyl Using a Novel Hapten and Monoclonal Antibody. *Foods* **2023**, *12*, 2764. https://doi.org/10.3390/foods12142764

Academic Editor: Soledad Cerutti

Received: 7 June 2023
Revised: 12 July 2023
Accepted: 14 July 2023
Published: 20 July 2023

1. Introduction

Due to their high efficiency and low toxicity [1–3], sulfonylurea-based herbicides are widely used to remove noxious weeds in maize, cereal, sugarcane, and tomato fields. HM (a sulfonylurea) is the active ingredient in many registered products in China. Currently, there are 14 registered manufacturers of HM pesticides in China, with a total of 22 registered formulations. These formulations are mainly registered for use in 10 crop fields, including winter wheat fields, tomato fields, sugarcane fields, sorghum fields, upland direct-seeded rice fields, paddy fields (direct-seeded), transplanted rice fields, summer corn fields, wheat fields, and corn fields (source: China Pesticide Information Network). The excessive or repeated use of herbicides without following proper scientific methods for weed control can also lead to serious phytotoxicity, directly impacting crop yields. Although HM is metabolized quickly in target crops, its metabolites easily leach to deep layers of the water table with rainwater and irrigation water, causing pollution of water resources, which can endanger human health. In addition, HM residues in water and soil have caused damage to nontarget crops such as soybean [4]. The potential impact of HM metabolites on other nontarget crops and aquatic ecosystems is still not fully understood.

Monitoring of HM and other sulfonylureas in food samples is typically accomplished using instrumental methods, such as liquid chromatography-tandem mass spectrometry (LC-MS/MS) [2,5], capillary electrophoresis (CE)-MS/MS [3], and ultra-pressure liquid chromatography (UPLC) [6]. While these methods are sensitive and accurate, they are time intensive and require operation by specialized personnel. Immunoassay, being a widely utilized method in the rapid detection of pesticides, small molecules, environmental pollutants, viruses, and hormones, offers certain advantages over instrumental analytical methods in terms of simplicity, speed, and portability. The basis of immunoassay lies in the design and preparation of highly specific haptens and antibodies, as well as the selection of appropriate immunodetection modes.

Hapten design considerations include the position of the active site group and the length of the connecting spacer arm, as well as the stability and solubility of the hapten [7–10]. Due to the instability of sulfonylureas, a common strategy for hapten synthesis involves adding a functional spacer arm to a stable metabolite of sulfonylurea [11–13].

For instance, the addition of a succinic acid spacer arm to stable metabolites of triasulfuron, metsulfuron-methyl, and chlorimuron-ethyl resulted in the production of monoclonal antibodies (mAbs) with enhanced sensitivity [11–13]. These antibodies were then applied in the development of highly sensitive indirect competitive ELISAs for target molecule detection. In a specific study, Schlaeppi et al. [11] prepared two triasulfuron haptens, one with a functional aminoalkyl group added to the triazine ring and the other with a functional succinic acid spacer arm added to the chloroethoxy sulfonamide moiety. The latter hapten generated a more sensitive mAb for triasulfuron. Similarly, Welzig et al. [12] generated metsulfuron-methyl rabbit antibodies using a hapten synthesized by adding a succinic acid C3-spacer arm to a phenylsulfonyl derivative of metsulfuron-methyl. Zhao et al. [13] included a succinic acid spacer arm in their chlorimuron-ethyl hapten, specifically added to the ethoxycarbonyl phenyl sulfonamide of its metabolite. This hapten enabled the generation of a highly sensitive mAb to chlorimuron-ethyl.

Due to the presence of the sulfonylurea bridge in the molecular structure of HM, the molecule has poor stability, resulting in high cost and low success rate in synthesizing the complete HM molecule as a hapten. This limitation has hindered the development of HM antibodies and the establishment of immunoassay. Therefore, the design of an appropriate hapten is crucial [14]. However, given its hydrolysis to a pyrazole-sulfonamide and pyrimidine amine, strategies similar to those mentioned above are expected to be needed for effective HM hapten design.

Magnetic nanoparticles (MNPs) have good stability, high operability, and excellent biocompatibility, making them broadly applicable to food safety detection, environmental treatments, biological medicine, etc. MNPs are generally employed in analyte separations and enrichment, as drug carriers [15,16], and as signal probes in magnetic lateral flow strips [17–19]. Due to their high separation and enrichment efficiency, a magnetic lateral flow immunoassay could provide high sensitivity detection of HM in water.

In this study, haptens of stable metabolites of HM with linkers tethered at alternative sites of the metabolite's structure were used to raise specific monoclonal antibodies to HM. The resulting monoclonal antibodies were used to establish a highly sensitive dcELISA for rapid detection of residual HM in registered food matrices, including tomato and maize. Additionally, we developed a sensitive lateral flow immunoassay based on 50 nm MNPs for the detection of HM in rice paddy water. These findings have significant implications for the efficient and accurate detection of HM residues and contribute to the development of safer agricultural practices.

2. Materials and Methods

2.1. Reagents, Buffers, and Instruments

Bovine serum albumin (BSA); ovalbumin (OVA); horseradish peroxidase (HRP); complete and incomplete Freud's adjuvant; poly-ethylene glycol (PEG, Mw = 1450); hypoxanthine, aminopterin, and thymidine (HAT) medium supplements; cell-freezing medium; dimethyl sulfoxide (DMSO); N-hydroxysulfosuccinimide sodium salt (sulfo-NHS); 1-(3-dimethylaminopropyl)-3-ethylcarbodiimide hydrochloride (EDC); o-phenylenediamine (OPD); and mouse monoclonal antibody isotyping reagents were purchased from Sigma-Aldrich (St. Louis, MO, USA). Succinic anhydride and 5-(aminosulfonyl)-3-chloro-1-methyl-1H-pyrazol, the 50 nm carboxyl magnetic beads (10 mg/mL), and 2-(N-morpholino) ethane-sulfonic acid (MES) were from Shanghai yuanye Bio-Technology Co., Ltd. (Shanghai, China). Raw tomato and maize samples were obtained from the Nankou Experimental Station of the Chinese Academy of Agricultural Sciences (Beijing, China), where herbicide treatments were not applied during the cultivation.

Buffers used in this study were made as previously described by Cui et al. [1]. Phosphate-buffered saline (PBS, 0.01 M, pH 7.4), carbonate-buffered saline (cbuffer; 0.5 M, pH 9.6), washing buffer (PBS containing 0.1% Tween-20), sample diluting buffer (PBSTG, PBS containing 0.1% Tween-20 and 0.5% gelatin), and substrate buffer (pH 7.4, 0.01 M citrate–phosphate) were used in sample dilution and ELISA experiments. The 10 mL substrate solution, containing 20 mg OPD and 4 µL of 30% hydrogen peroxide, was prepared freshly before use.

2.2. Synthesis of the HM Hapten

According to the synthetic procedure in Figure 1A, the HM hapten was synthesized by reacting pyrazole sulfonamide (a HM metabolite) and succinic anhydride. A solution of 5-(aminosulfonyl)-3-chloro-1-methyl-1H- pyrazol (1 g, 4 mmol) and succinic anhydride (0.4 g, 4 mmol) was slowly treated with DBU, and the mixture was stirred at room temperature for 3 h. The reaction mixture was acidified to pH 2 with 2 M hydrochloric acid and extracted three times with 50 mL ethyl acetate. The organic layers were dried over anhydrous Na_2SO_4, filtered, and concentrated. The buffer residues were purified with a chromatographic silica gel column using dichloromethane/methyl alcohol (10:1) as the eluent, which yielded HM hapten (0.9 g, 66.6%). The high-resolution mass spectrometry (HRMS) and nuclear magnetic resonance (NMR) spectra of the HM hapten are shown in Figures S1 and S2. HRMS was conducted using a quadrupole time-of-flight mass spectrometer (QTOF MS), and the accurate mass $[M\text{-}H]^- = 352.0019$ was detected, which corresponded to $C_{10}H_{12}ClN_3O_7S$, which has an exact mass of $[M\text{-}H]^- = 352.0006$. ^1H NMR (400 MHz, DMSO) resulted in the following shifts: δ 12.32 (b2, H, COOH), δ 4.13 (s, 3H, NMe), 3.84 (s, 3H, OMe), 2.53 (dd, J = 12.6, 6.3 Hz, 2H, CH2), and 2.42 (t, J = 6.1 Hz, 2H, CH2); ^{13}C NMR (400 MHz, DMSO): δ 174.36, 134.69, 129.27, 121.91, 121.71, 119.75, 87.75, 27.77, 25.72, and 24.17 ppm.

2.3. Preparation of the Protein-Hapten Conjugate

The HM hapten was conjugated with carrier proteins, bovine serum albumin (BSA), ovalbumin (OVA), and horseradish peroxidase (HRP) using the carbodiimide cross-linker method as previously described [20]. Hapten to protein conjugation was carried out following the synthetic route illustrated in Figure 1B. Briefly, the HM hapten, EDC, and sulfo-NHS were dissolved in 500 µL DMSO at a ratio of n(hapten):n(EDC):n(sulfo-NHS) = 1:1.25:1.25 by stirring for 4 h at room temperature. The reaction mixture was then centrifuged at $8452\times g$ for 5 min to remove the precipitate. The activated hapten was added dropwise to the protein solution (0.01 M PBS) at a ratio of n(protein):n(hapten) = 1:30 and stirred at 4 °C for 4 h. The mixture was dialyzed against 4 × 4 L PBS at 4 °C over 20 h. After dialysis, BSA-hapten, OVA-hapten, and HRP hapten conjugates were stored at −20 °C. The molar ratio of the protein-hapten was tested on a Bruker Autoflex III matrix-assisted laser desorption/ionization time of flight mass spectrometry (MALDI-TOF-MS; Bruker Corporation, Billerica, MA, USA).

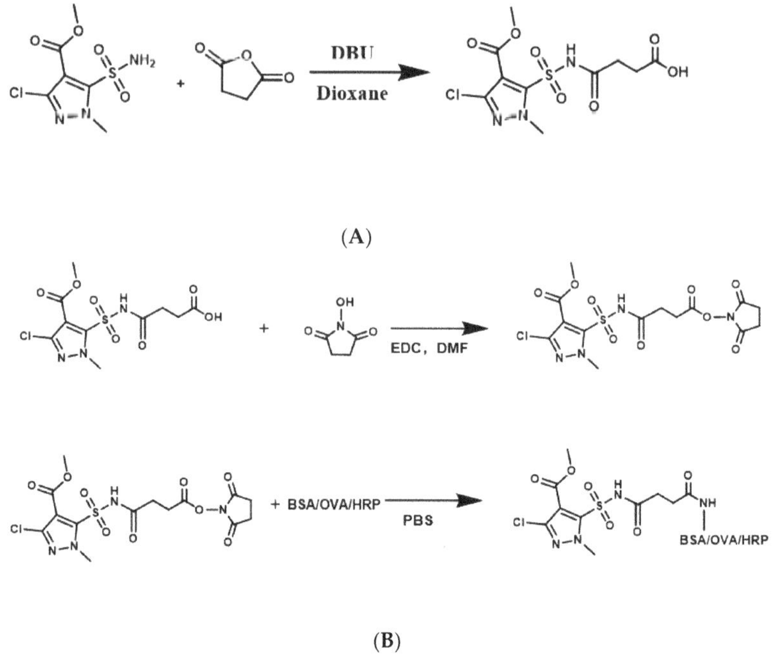

(A)

(B)

Figure 1. Schematics of HM hapten synthesis (**A**) and HM antigen preparation (**B**).

2.4. Immunization and Monoclonal Antibody Production

Animal immunization, cell fusion, and monoclonal antibody production protocols were the same as was previously described [20–22]. The ethical approval for our animal experiments was granted by the Experimental Animal Welfare and Ethical Committee of Institute of Quality Standards and Testing Technology for Agro-Products, Chinese Academy of Agricultural Sciences (IQSTAP-2021-05). Female BALB/C mice (6–8 weeks old) were immunized with the BSA-hapten conjugate by intraperitoneal and subcutaneous injection. A 100 µg amount of the BSA-hapten conjugate was dissolved in 100 µL PBS and emulsified with 100 µL complete Freund's adjuvant. Subsequent doses were given using incomplete Freund's adjuvant every two weeks. The antisera were collected after the third injection to test the titer and target recognition ability of the antibody by checkboard icELISA. Positive mouse spleen cells were fused with SP2/0 myeloma cells at a ratio of 10:1 using PEG with MW = 1450. The fused cells were cultured in HAT selection medium. Ten days after fusion, cell supernatants were screened for their response to HM using icELISA. Individuals exhibiting a strong positive response to HM and high levels of inhibition were isolated using the limiting dilution method. After one or two subclones, the monoclonal hybridoma cells were selected and expanded in a 6-well cell plate. Next, 106 of the selected monoclonal hybridoma cells were injected into BALB/c mice to produce ascites. The ascites were purified via ammonium sulfate precipitation, dialyzed against 0.01 M PBS to obtain a monoclonal antibody, and stored at $-20\,^{\circ}\text{C}$ after freeze-drying.

The isotype of the obtained monoclonal antibody was detected using a mouse monoclonal antibody isotyping kit. The sensitivity (50% inhibitory concentration, IC50) and limit of detection (20% inhibitory concentration, IC20) were calculated from a standard curve. The specificity of the monoclonal antibody was determined by calculating the cross-reactivity (CR) of the monoclonal antibody to five other sulfonylurea herbicides. CR was calculated as the IC50 of HM IC50 of analogs.

2.5. Establishment and Optimization of the dcELISA

The direct and competitive ELISAs were developed using the above-described monoclonal antibody and the coating antigen HRP-hapten conjugate. The format of dcELISA is shown in Figure 2. Briefly, polystyrene microplate wells were coated with a monoclonal antibody in PBS (pH = 7.4, 100 µL/well) for 3 h at 4 °C; then, the plate was washed three times with PBSTG. The HM standard (50 µL/well) and HRP-hapten conjugate (50 µL/well) were diluted in PBSTG and added into wells sequentially. The plate was incubated for 15 min at 4 °C and washed three times with PBSTG. After passing the plate three times, the substrate solution was added, and the reaction was stopped as mentioned for the icELISA above.

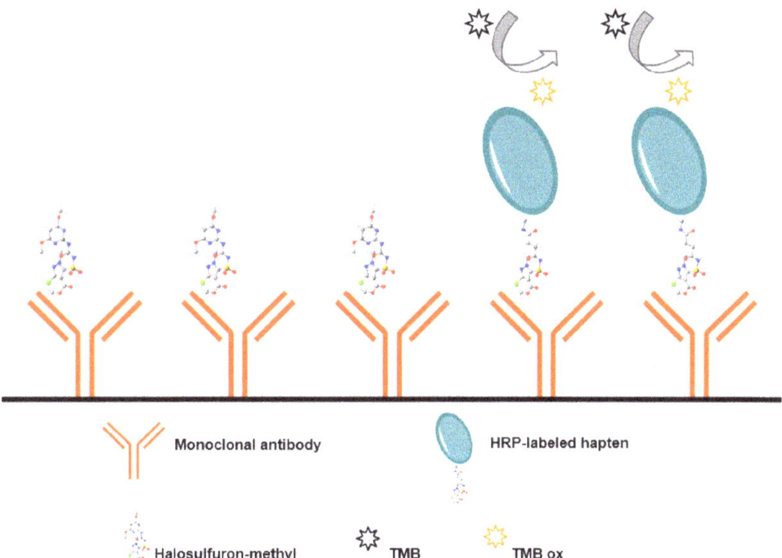

Figure 2. Schematic of dcELISA.

The buffer pH, ionic strength, and organic solvent were optimized for the dcELISA. The sample diluting buffer was evaluated from pH = 5.5 to pH = 9.5. The ionic strength effect on the sample diluting buffer was estimated by changing the Na^+ concentration from 0.05 M to 0.2 M. The organic solvent content effect was tested using different concentrations (5%, 10%, 15%, and 20%) of methanol and acetonitrile in sample diluting buffer, respectively. The results were evaluated based on the IC50 value and the coefficient of variation (R^2) of the linear equation. The absorbance was measured at 490 nm on an Infinite M200 Pro microplate reader (Tecan Group Ltd., Männedorf, Switzerland).

2.6. Functionalization of MNPs with Antibodies

Conjugation of the magnetic nanoparticles with antibodies was performed according to a previous report [13]. In brief, a 50 µL aliquot of carboxyl magnetic beads (50 nm) was transferred to a "LoBind protein" microcentrifuge tube. Then, 50 µL of 10 mg/mL EDC and 150 µL of 10 mg/mL NHS were added. After mixing at room temperature for 20 min, the supernatant (containing PBST) was removed with a magnetic separator, and then the magnetic beads were resuspended in coupling buffer. For conjugation, 25 µg HM mAbs 1A91H11 were added to the magnetic beads, and the free antibodies were washed with PBST three times. After mixing for 2 h at ambient temperature, the MNPs/antibodies conjugation was blocked with blocking buffer for 20 min and washed with PBST three times. Finally, the precipitate was re-suspended in PBS solution (0.01 mol/L, pH = 7.4) and stored at 4 °C until use.

2.7. Establishment of the MLFIA

A mixture of 10 µL MNPs and 1 mL sample extract was added to a microcentrifuge tube and mixed at room temperature for 30 min. After removing the supernatant using the magnetic separator, the MNPs were redispersed in 100 µL 0.01 M PBS (pH = 7.4).

The lateral flow strip was assembled by sequentially pasting the adsorbent pad, nitrocellulose membrane, and sample pad onto a polyvinyl chloride (PVC) plate. The adsorbent pad and sample pad overlapped 1 and 2 mm with the nitrocellulose membrane, respectively. The nitrocellulose membrane was cut into 4 mm wide strips with a guillotine cutter. Then, the HM hapten-BSA conjugate and goat antimouse IgG were sprayed onto the nitrocellulose membrane to form test line (T) and a control line, respectively. Afterwards, the NC membrane with the test line and the control line was dried at 37 °C for 1 h.

2.8. Sample Extraction and Analysis

The extraction method refers to a reference with minor modifications [1]. The homogenized tomato sample (5 g) was weighed and extracted with 10 mL acetonitrile. After stirring at room temperature for 5 min, 1 g NaCl and 4 g anhydrous $MgSO_4$ were added to the sample. The mixture was stirred for 3 min and then centrifuged at 5000 rpm for 5 min. The supernatant (1 mL) was further purified by adding 50 mg of C18 resin. After being centrifuged at 500 rpm for 5 min, the extract was dried under a nitrogen stream and resuspend by sample dilution buffer for further dcELISA analysis. For LC-MS/MS analysis, the extract was filtered through a 0.22 µM membrane and detected.

The paddy water (5 mL) was taken and filtered through a 0.22 µM membrane directly for MLFIA and LC-MS/MS analysis.

3. Results and Discussion

3.1. Characterization of the HM Hapten and Its Bioconjugates

The sulfonylurea bridge of HM is unstable, being readily hydrolyzed through bridge contraction and rearrangement under strong pH conditions [14], making synthesis of the hapten with the intact molecule infeasible. Simon et al. [23] and Zhao et al. [24] reported on the synthesis of haptens for metsulfuron-methyl and chlorimuron-ethyl, respectively, to produce antibodies. In those studies, the haptens contained half moieties of the herbicides and a succinic acid spacer arm that was conjugated to carrier protein, which was used to and immunize mice to produce high-affinity antibodies. We similarly designed a HM hapten and produced specific antibodies to HM (Figure 1A). By introducing a spacer arm through a C-N single bond at a sulfonamide position of 5-(aminosulfonyl)-3-chloro-1-methyl-1H-pyrazol moiety, this hapten was able to preserve the structure of HM with minimal alteration to the original framework.

Three HM bioconjugates, including hapten-BSA, hapten-OVA, and hapten-HRP, were prepared and used in the present study. First, the carboxylate group of the hapten was activated by N-hydroxysuccinimidyl ester, forming an active derivative. This active derivative reacted with the amino group on the carrier protein to form a bioconjugate, and the precise molar ratio (MR) of the hapten to protein was detected using MALDI-TOF-MS and calculated as follows: MR = (the mass of conjugate − the mass of protein)/mass of hapten. Mass spectra of the protein standards and bioconjugates are shown in Figure S3. The MRs were estimated to be 7:1, 2:1, and 1:1 for BSA-hapten, OVA-hapten, and HRP-hapten, respectively.

3.2. Antibody Generation and Characterization

Novel monoclonal antibodies for HM were produced from mice immunized with the BSA-hapten conjugate. The mouse with the highest titer and the best inhibition to HM in checkerboard icELISA was used for screening monoclonal hybridoma cells. After cloning, a limiting dilution assay was used to identify a positive monoclonal hybridoma, clone 1A91H11, that secreted antibodies against HM. Clone 1A91H11 was then cultivated and used to produce monoclonal antibodies. According to a checkerboard titration, the optimal dilution factor of the coating antigen OVA-hapten conjugate (1 mg/mL) and monoclonal antibody 1A91H11 (1 mg/mL) was 1.6×10^5 times for icELISA. Figure S4 shows the standard inhibition curve of HM. The IC50 value and detection range were 16.5 ng/mL and 8.1 ng/mL–44.9 ng/mL (inhibition rate between 20–80%).

To evaluate the specificity of monoclonal antibody 1A91H11, cross-reactivity (CR%) of the antibody with several sulfonyl ureas, including 5-(aminosulfonyl)-3-chloro-1-methyl-1H-pyrazol, pyrazosulfuron-ethyl, nicosulfuron, chlorsulfuron, ethoxysulfuron, and chlorimuron-ethyl, were evaluated using icELISA. Except for the 0.06% CR with 5-(aminosulfonyl)-3-chloro-1-methyl-1H- pyrazol, MAb 1A91H11 showed no cross-reactivity with the tested sulfonylurea herbicides (Table 1). This indicated the validity of the novel hapten and the acceptable specificity of the antibody for accurate HM analysis.

3.3. Optimization of dcELISA Conditions

dcELISA is an immobilized antibody assay based on the competition between an enzyme-labeled hapten and an unknown amount of analyte for the immobilized antibody. dcELSIA is more sensitive and time-saving than icELISA [25]. To optimize the HM dcELISA, influencing factors, such as the optimal ratio of coating monoclonal antibody and hapten-HRP, pH value, and the diluting buffer NaCl concentration, were determined. According to the checkerboard dcELISA results, the dilution ratios of coating monoclonal antibody 1A91H11 (1 mg/mL) and hapten-HRP conjugate (1 mg/mL) were 1:200 and 1:3200. As shown in Figure 3, among the tested diluting buffer pH values (5.5, 6.5, 7.5, 8.5, and 9.5), the lowest IC50 (1.5 ng/mL) was achieved at pH 6.5. The ionic strength had little effect on the IC50 of the dcELISA when the NaCl concentration ranged from 0.05 to 0.2 M. When the diluting buffer contained 5% methanol (v/v), the sensitivity of dcELISA significantly decreased (IC50 42.5 ng/mL), most likely due to inhibition of the HRP-labeled hapten and the interaction between the antigen and antibody. Thus, methanol should be avoided in the extraction. As the acetonitrile concentration was increased from 5% to 20%, the IC50 values were observed to fall within the range of 10.15 ng/mL–26.91 ng/mL, indicating a minimal influence of acetonitrile on the dcELISA. This suggests that acetonitrile can be considered as a suitable extraction solvent when using the dcELISA method for sample detection.

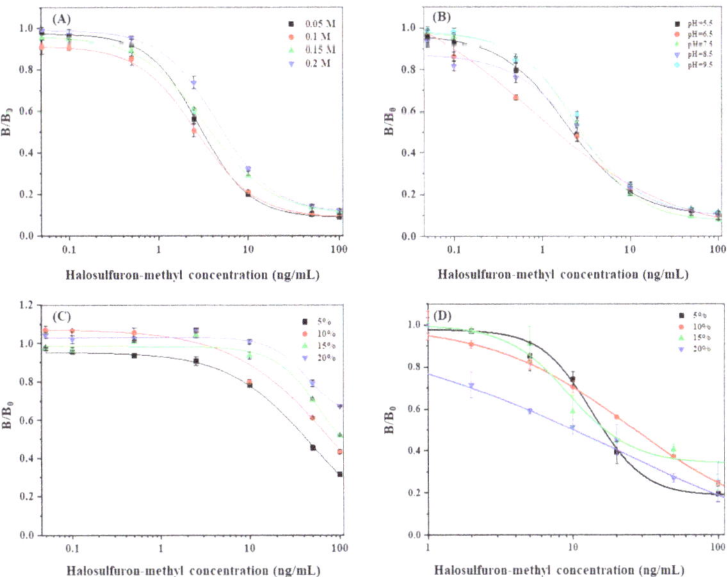

Figure 3. Optimization of dcELISA: (**A**) pH; (**B**) Na$^+$ concentration; (**C**) Methanol concentration; (**D**) Acetonitrile concentration.

Table 1. Cross reactivity of mAb 1A91H11 with HM and other sulfonylureas.

Analytes and Structures		IC50 (ng/mL)	Cross Reactivity
HM	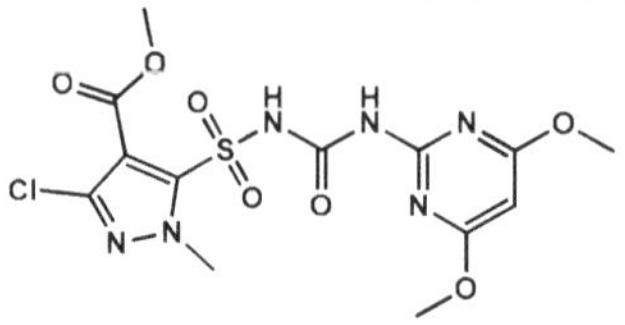	28.89	100%
HM hapten	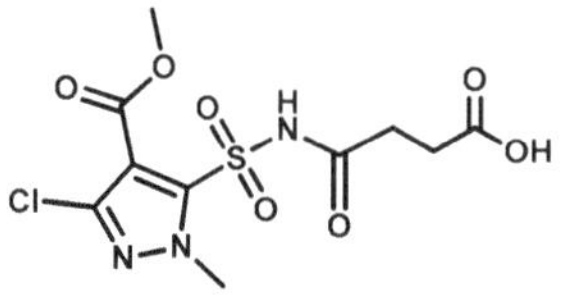	43.5 [a]	0.06% [a]
Nicosulfuron	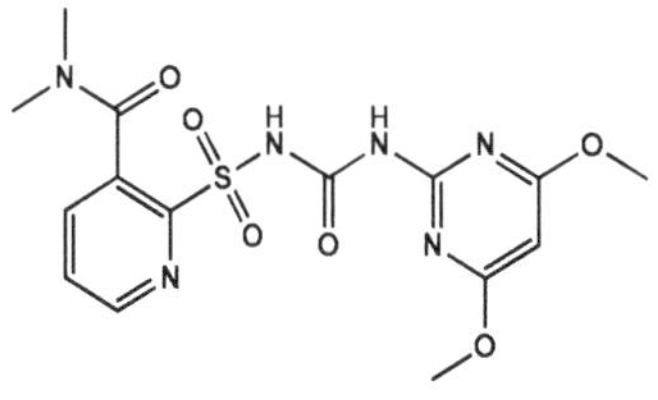	>20,000	<0.02%
Chlosulfuron	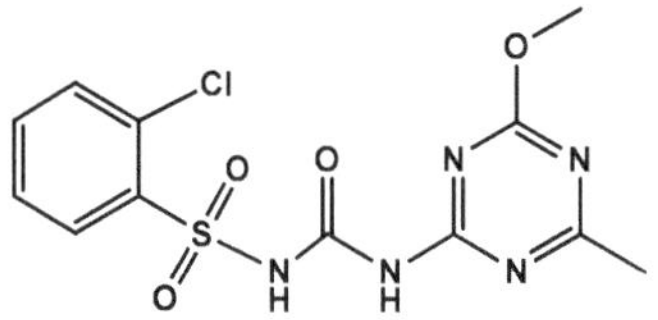	>20,000	<0.02%
Ethosulfuron	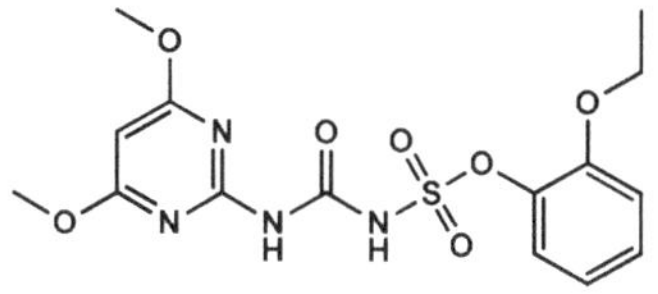	>20,000	<0.02%
Chlorimuron-ethyl	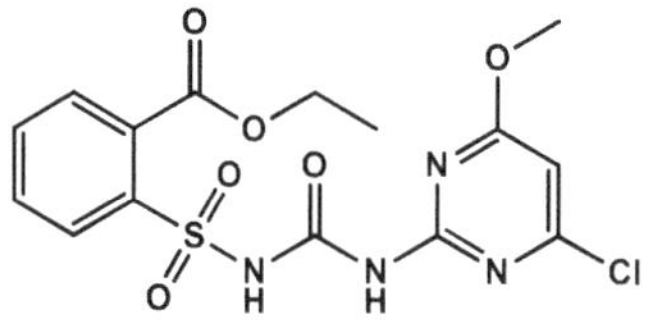	>20,000	<0.02%
Pyrazosulfuron-ethyl	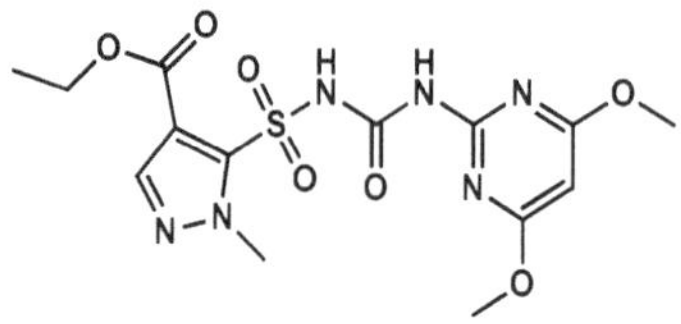	>20,000	<0.02%

[a] µg/kg.

The inhibition curve of dcELISA was obtained under the determined optimal conditions, pH 6.5, 0.01 M NaCl, and without methanol, resulting in an IC50 value of 1.5×10^{-3} mg/kg and working range of 0.7×10^{-3} mg/kg–10.7×10^{-3} mg/kg (inhibition rate between 20–80%) as shown in Figure 4. The sensitivity (IC50 = 1.5×10^{-3} mg/kg) of the dcELISA method for detecting HM residues was lower than the LOD of 2×10^{-3} mg/kg achieved by the CE-MS/MS method and similar to the LOQ of 1×10^{-3} mg/kg reported for the LC-MS/MS method [2,3].

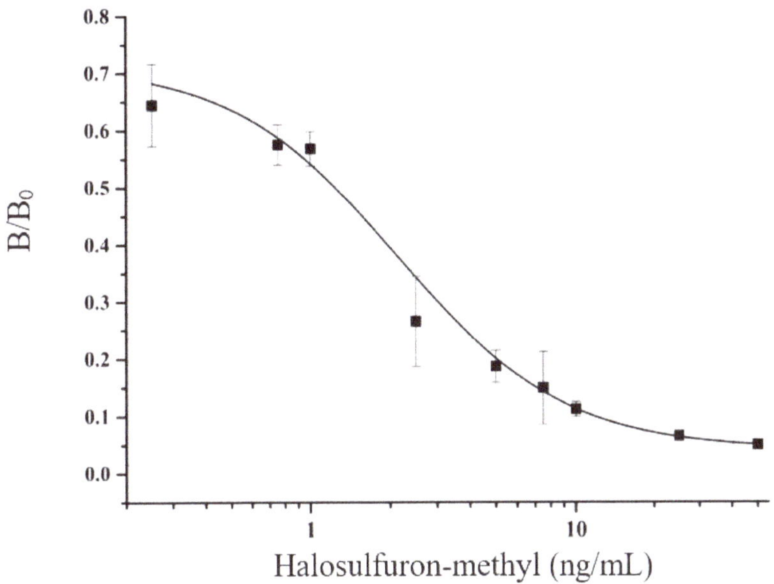

Figure 4. dcELISA calibration curve of HM.

3.4. Optimization of the MLFIA

The ratio of EDC to NHS, amount of HM antibodies, pH of the magnetic beads–antibodies coupling reaction, and the concentrations of hapten-BSA and goat–antimouse antibody applied to the NC membrane were systematically investigated to improve the sensitivity of the MLFIA. The format of MLFIA is shown in Figure 5.

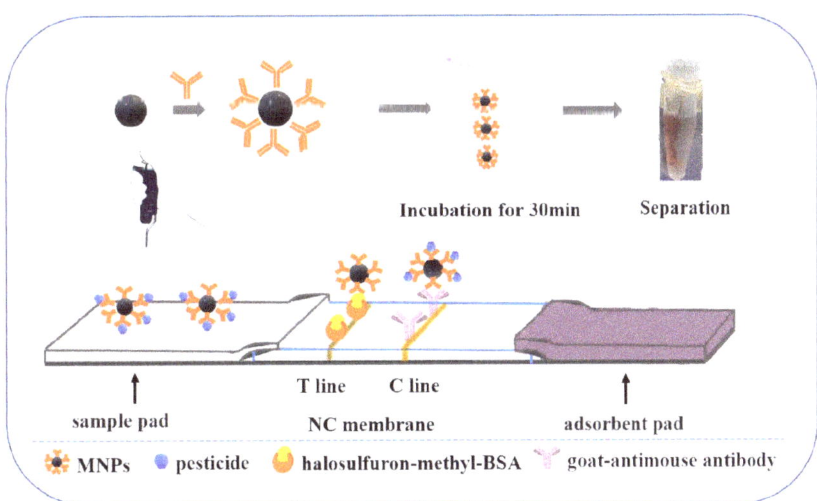

Figure 5. Illustration of a magnetic lateral flow immunoassay.

3.4.1. The EDC/NHS Ratio

Complementary interactions between EDC and NHS can increase the stability and yield of the conjugation reaction. Therefore, different ratios of EDC/NHS (from 1:1 to 1:4) were evaluated. As shown in Figure 6, the colors of both the T and C lines deepened as the EDC/NHS ratio increased (Figure 6A). However, no obvious difference was observed between 1:3 and 1:4. On the basis of our experimental results, 1:3 was chosen as the EDC/NHS ratio for the MLFIA.

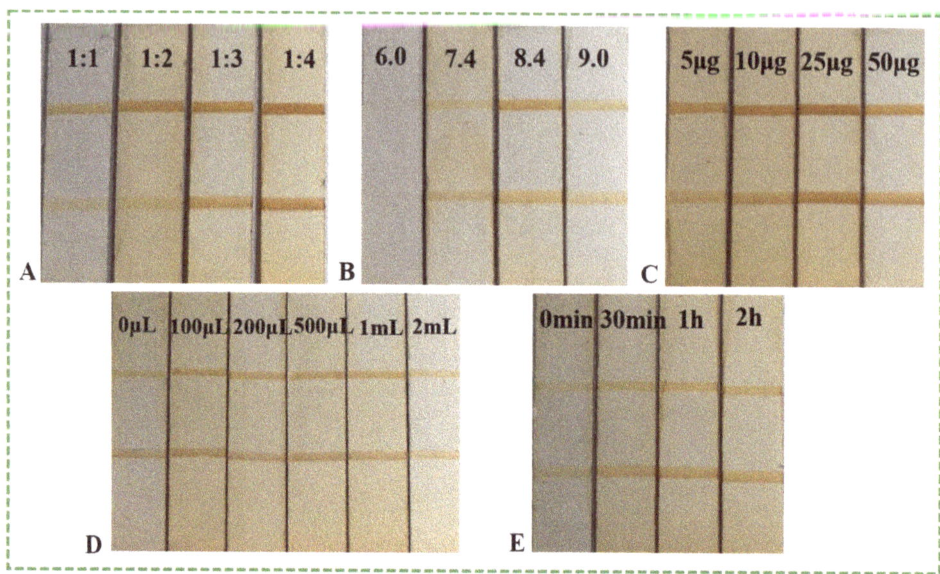

Figure 6. Optimization of MLFIA: (**A**) EDC/NHS ratio; (**B**) pH value; (**C**) Antibody amount; (**D**) Enrichment volume; (**E**) Enrichment time.

3.4.2. HM Antibodies Content and pH of the Conjugation Solution

Four different amounts of HM antibody 1A91H11 (5 µg, 10 µg, 25 µg, and 50 µg) were tested. As shown in Figure 6C, the darkest of T and C lines were obtained when 25 µg were added. Considering the pH can affect the coupling efficiency, the influence of pH from 6.0 to 9.0 was evaluated. The darkest T and C lines were obtained at pH 8.4 (Figure 6B). Therefore, 25 µg of HM antibody and pH 8.4 were selected as the optimal conditions of the conjugation solution.

3.4.3. Optimization of MNP Probe Enrichment Time and Volume

When added to LFIA, MNPs improve the detection limit by acting as color reagents and by providing an enrichment effect. To obtain better immunoassay performance, MNP probe enrichment time and volume were optimized. Ten µL MNP probes were added into 10 ng/mL HM at different volumes (100 µL, 200 µL, 500 µL, 1 mL, and 2 mL). As displayed in Figure 6D, the colors of both the T and C lines deepened with the addition of 200 µL, remained the same from 200 µL to 1 mL, and lightened at 2 mL. Subsequently, the mixture of MNP probes and HM standard were stirred for 15 min, 30 min, 1 h, and 2 h. As revealed in Figure 6E, colors of both the T and C lines increased from 15 min to 30 min and remained the same after 30 min. Therefore, 1 mL and 30 min were chosen as the optimal enrichment volume and time for the following experiments.

3.5. Evaluation and Specificity of MLFIA and dcELISA

Compared with dcELISA (IC50 1.5 ng/mL), MLFIA has the advantage of high sensitivity due to the enrichment effect and reduction of matrix effects. Serial dilutions of the extraction solution for MLFIA ranging from 0.01 to 5 ng/mL were tested, and standard curves of the MLFIA for HM were established under the optimal assay conditions

(Figure 7A). The assay had an IC50 value of 0.21 ng/mL and good linear fitting ($R^2 \geq 0.95$) with concentrations ranging from 0.03 to 1.5 ng/mL (inhibition rate between 20 and 80%).

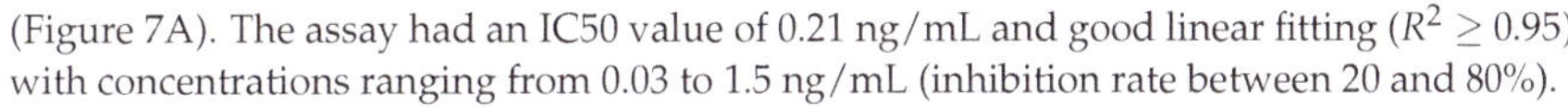

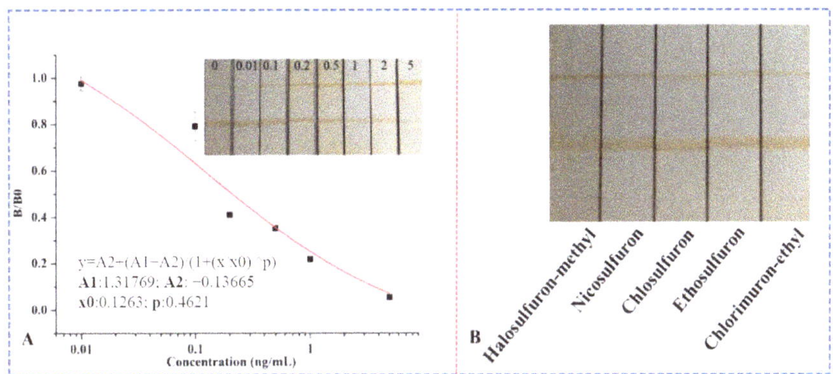

Figure 7. Standard curve for HM analysis (**A**) and specificity results (**B**) of MLFIA.

To evaluate the specificity of MLFIA for the detection of HM, 2 ng/mL of HM and 20 ng/mL of pyrazosulfuron-ethyl, nicosulfuron, chlorsulfuron, ethoxysulfuron, and chlorimuron-ethyl standard solutions were detected using MLFIA, in which the color intensity of T and C lines was measured using a test strip reader. The results are shown in Figure 7B. The T/C values of the blank solution and the analog standards were higher than the HM standard, suggesting good selectivity of the MLFIA, which was consistent with the mAb specificity.

3.6. Food Sample Analysis

HM is widely registered for use in tomato and maize cultivation in China, which poses the risk that its residue could be present in those foods. Tomato and maize samples were spiked with 0.025 mg/kg, 0.05 mg/kg and 0.1 mg/kg HM. The average recoveries were detected using dcELISA and LC-MS/MS. According to results shown in Table 2, the average recoveries of HM using dcELISA were 78.9–87.9% and 103.0–107.4% in tomato and maize with RSDs of 1.1–6.8% and 2.7–6.4%, respectively. The average recoveries of HM using LC-MS were 77.2–87.6% and 92.1–93.0% in tomato and maize with RSDs of 1.2–2.6% and 6.0–7.7%, respectively. The results of dcELISA were in agreement with those detected using LCMS/MS, which strengthens the reliability of dcELISA for the analysis of HM residue in tomato and maize matrices.

Table 2. Comparison between dcELISA and LC-MS/MS for halosulfuron recovery in tomato and maize.

Matrix	Spiked Halosulfuron Methyl (mg/kg)	Average Recovery and RSD (%)			
		dcELISA	RSD	LC-MS/MS	RSD
Tomato	0.025	78.9 ± 0.02	9.1	77.2 ± 0.9	1.2
	0.05	84.4 ± 0.01	1.1	87.2 ± 2.3	2.6
	0.1	87.9 ± 0.01	6.8	87.6 ± 1.3	1.8
Maize	0.025	107.4 ± 5.6	6.4	92.8 ± 8.2	6.0
	0.05	105.5 ± 7.3	7.7	93.0 ± 8.5	7.7
	0.1	103.0 ± 2.6	2.7	92.1 ± 6.0	7.6

Similarly, paddy water was spiked with 0.025, 0.05, and 0.1 mg/kg HM standard. The results are presented in Table 3. The average recoveries and RSDs of the LC-MS/MS ranged from 93.5 to 98.9% and 1.8 to 2.7%, while the average recoveries and RSDs of the developed MLFIA ranged from 81.5 to 92.5% and 5.4 to 9.7%, which generally satisfied the requirements of trace detection. Regarding paddy water, the developed MLFIA showed acceptable average recoveries and RSDs. The developed MLFIA method may be suitable for the rapid analysis of HM residue in paddy water samples.

Table 3. Comparison between MLFIA and LC-MS/MS analysis for halosulfuron recovery in paddy water.

Matrix	Spiked Halosulfuron Methyl (mg/kg)	Average Recovery and RSD (%)			
		MLFIA	**RSD**	**LC-MS/MS**	**RSD**
Paddy water	0.025	81.5	8.3	93.5	1.8
	0.05	92.5	9.7	98.9	2.3
	0.1	89.3	5.4	96.5	2.7

4. Conclusions

In the present research, the design of a novel HM hapten followed a similar method to that of other sulfonylurea herbicides, such as triasulfuron [11], chlorsulfuron [26], and chlorimuron [13]; i.e., one step synthesis of the hapten was carried out with the half moiety of HM. Unlike other studies, the present hapten introduced a carboxyl group to the pyrazole-sulfonamide moiety of halosulfuron with a four-carbon length, which enabled sufficient exposure of the carrier-protein-attached hapten to the immune system [10]. Due to this, the resulting Mab 1A91H11 was highly sensitive to the target analyte. A highly sensitive dcELISA using Mab 1A91H11 and hapten-labeled HRP was developed, optimized, and validated for the determination of HM in tomato and maize samples. The dcELISA developed here, with an IC50 value of 1.5 ng/mL, was more convenient and 10 times more sensitive than the typically employed indirect competitive assay to detect small molecules. This dcELISA was optimized to be a rapid and sensitive alternative tool to detect HM in tomato and maize samples. For the convenience of field detection, therefore, a magnetic-beads-based immunoassay with higher sensitivity and shorter detection time was developed. Magnetic beads played the role of concentration, enrichment, and signal detection. The MLFIA possessed an IC50 value of 0.21 ng/mL, which was 10 times lower than the dcELISA. MLFIA provides a development direction worthy of attention for rapid and highly sensitive detection of HM and other pesticide residues in food and the environment.

Supplementary Materials: The following supporting information can be downloaded at https://www.mdpi.com/article/10.3390/foods12142764/s1. Figure S1: HRMS of the halosulfuron-methyl hapten. Figure S2: ^{1}H NMR (a) and ^{13}C NMR (b) spectra of the halosulfuron-methyl hapten. Figure S3: MALDI-TOF MS of halosulfuron-methyl antigens: (A) BSA vs. the BSA-hapten conjugate; (B) OVA vs. the OVA-hapten conjugate; (C) HRP vs. the HRP-hapten conjugate. Figure S4: The icELISA calibration curve of halosulfuron-methyl.

Author Contributions: Conceptualization, Y.Y., Z.C. and J.W.; Data curation, X.C. and T.L.; Formal analysis, Y.Y. and X.C.; Funding acquisition, Z.C. and J.W.; Investigation, X.C.; Methodology, Y.Y., H.L., L.P. and T.L.; Supervision, Z.C.; Writing—original draft, Y.Y.; Writing—review and editing, H.L. All authors have read and agreed to the published version of the manuscript.

Funding: This work was supported by funding from the National Natural Science Foundation of China (32202177) and Agricultural Science and Technology Innovation Program of Chinese Academy of Agricultural Sciences (CAAS-ZDRW202011).

Data Availability Statement: The data used to support the findings of this study can be made available by the corresponding author upon request.

Conflicts of Interest: The authors declare no conflict of interest.

References

1. Ying, Y.; Cao, Z.; Li, H.; He, J.; Zheng, L.; Jin, M.; Wang, J. An optimized LC-MS/MS workflow for evaluating storage stability of fluroxypyr and halosulfuron-methyl in maize samples. *J. Environ. Sci. Health B* **2021**, *56*, 64–72. [CrossRef] [PubMed]
2. Ni, Y.; Yang, H.; Zhang, H.; He, Q.; Huang, S.; Qin, M.; Chai, S.; Gao, H.; Ma, Y. Analysis of four sulfonylurea herbicides in cereals using modified Quick, Easy, Cheap, Effective, Rugged, and Safe sample preparation method coupled with liquid chromatography-tandem mass spectrometry. *J. Chromatogr. A* **2018**, *1537*, 27–34. [CrossRef] [PubMed]
3. Daniel, D.; dos Santos, V.B.; Vidal, D.T.R.; do Lago, C.L. Determination of halosulfuron-methyl herbicide in sugarcane juice and tomato by capillary electrophoresis-tandem mass spectrometry. *Food Chem.* **2015**, *175*, 82–84. [CrossRef] [PubMed]
4. Li, Y.; Zhang, Q.; Yu, Y.; Li, X.; Tan, H. Integrated proteomics, metabolomics and physiological analyses for dissecting the toxic effects of halosulfuron-methyl on soybean seedlings (Glycine max merr.). *Plant Physiol. Biochem.* **2020**, *157*, 303–315. [CrossRef] [PubMed]
5. Zhong, M.; Wang, T.; Dong, B.; Hu, J. QuEChERS-based study on residue determination and dissipation of three herbicides in corn fields using HPLC-MS/MS. *Toxicol. Environ. Chem.* **2015**, *98*, 216–225. [CrossRef]
6. Devi, R.; Duhan, A.; Punia, S.S.; Yadav, D.B. Degradation dynamics of halosulfuron-methyl in two textured soils. *Bull. Environ. Contam. Toxicol.* **2019**, *102*, 246–251. [CrossRef]
7. Ceballos-Alcantarilla, E.; Agulló, C.; Abad-Fuentes, A.; Abad-Somovilla, A.; Mercader, J.V. Rational design of a fluopyram hapten and preparation of bioconjugates and antibodies for immunoanalysis. *RSC Adv.* **2015**, *5*, 51337–51341. [CrossRef]
8. Goodrow, M.H.; Sanborn, J.R.; Stoutamire, D.W.; Gee, S.J.; Hammock, B.D. Strategies for immunoassay hapten design. In *ACS Symposium Series; Immunoanalysis of Agrochemicals: Emerging Technologies*; Nelson, J.O., Karu, A.E., Wong, R.B., Eds.; American Chemical Society: Washington, DC, USA, 1995; Volume 586, pp. 119–139.
9. Kim, Y.J.; Cho, Y.A.; Lee, H.-S.; Lee, Y.T.; Gee, S.J.; Hammock, B.D. Synthesis of haptens for immunoassay of organophosphorus pesticides and effect of heterology in hapten spacer arm length on immunoassay sensitivity. *Anal. Chim. Acta* **2002**, *475*, 85–96. [CrossRef]
10. Mercader, J.V.; Suarez-Pantaleon, C.; Agullo, C.; Abad-Somovilla, A.; Abad-Fuentes, A. Production and characterization of monoclonal antibodies specific to the strobilurin pesticide pyraclostrobin. *J. Agric. Food Chem.* **2008**, *56*, 7682–7690. [CrossRef]
11. Schlaeppi, J.M.A.; Meyer, W.; Ramsteiner, K.A. Determination of Triasulfuron in Soil by Monoclonal Antibody-Based Enzyme Immunoassay. *J. Agrlc. Food Chem.* **1992**, *40*, 1093–1098. [CrossRef]
12. Welzig, E.; Pichler, H.; Krska, R.; Knopp, D.; Niessner, R. Development of an Enzyme Immunoassay for the Determination of the Herbicide Metsulfuron-Methyl Based on Chicken Egg Yolk Antibodies. *Int. J. Environ. Anal. Chem.* **2000**, *78*, 279–288. [CrossRef]
13. Zhao, J.; Yi, G.-X.; He, S.-P.; Wang, B.-M.; Yu, C.-X.; Li, G.; Zhai, Z.-X.; Li, Z.-H.; Li, Q.X. Development of a Monoclonal Antibody-Based Enzyme-Linked Immunosorbent Assay for the Herbicide Chlorimuron-ethyl. *J. Agric. Food Chem.* **2006**, *5*, 4948–4953. [CrossRef] [PubMed]
14. Zheng, W.; Yates, S.R.; Papiernik, S.K. Transformation kinetics and mechanism of the sulfonylurea herbicides pyrazosulfuron ethyl and halosulfuron methyl in aqueous solutions. *J. Agric. Food Chem.* **2008**, *56*, 7367–7372. [CrossRef] [PubMed]
15. Xie, J.; Liu, G.; Eden, H.S.; Ai, H.; Chen, X. Surface-Engineered Magnetic Nanoparticle Platforms for Cancer Imaging and Therapy. *Acc. Chem. Res.* **2011**, *44*, 883–892. [CrossRef]
16. Chen, M.-L.; He, Y.-J.; Chen, X.-W.; Wang, J.-H. Quantum dots conjugated with Fe3O4-filled carbon nanotubes for cancer-targeted imaging and magnetically guided drug delivery. *Langmuir* **2012**, *28*, 16469–16476. [CrossRef]
17. Wu, J.; Dong, M.; Zhang, C.; Wang, Y.; Xie, M.; Chen, Y. Magnetic Lateral Flow Strip for the Detection of Cocaine in Urine by Naked Eyes and Smart Phone Camera. *Sensors* **2017**, *17*, 1286. [CrossRef]
18. Pang, L.; Quan, H.; Sun, Y.; Wang, P.; Ma, D.; Mu, P.; Chai, T.; Zhang, Y.; Hammock, B.D. A rapid competitive ELISA assay of Okadaic acid level based on epoxy-functionalized magnetic beads. *Food Agric. Immunol.* **2019**, *30*, 1286–1302. [CrossRef]
19. Hu, Y.; Shen, G.; Zhu, H.; Jiang, G. A Class-Specific Enzyme-Linked Immunosorbent Assay Based on Magnetic Particles for Multiresidue Organophosphorus Pesticides. *J. Agric. Food Chem.* **2010**, *58*, 2801–2806. [CrossRef]
20. Cui, Y.; Cao, Z.; Guo, S.; Zhang, W.; Tan, G.; Li, Z.; Wang, B. Hapten Synthesis and Monoclonal Antibody-Based Immunoassay Development for the Analysis of Thidiazuron. *J. Plant Growth Regul.* **2015**, *35*, 357–365. [CrossRef]
21. Cao, Z.; Zhao, H.; Cui, Y.; Zhang, L.; Tan, G.; Wang, B.; Li, Q.X. Development of a sensitive monoclonal antibody-based enzyme-linked immunosorbent assay for the analysis of paclobutrazol residue in wheat kernel. *J. Agric. Food Chem.* **2014**, *62*, 1826–1831. [CrossRef]
22. Zhang, R.; Liu, K.; Cui, Y.; Zhang, W.; He, L.; Guo, S.; Chen, Y.; Li, Q.X.; Liu, S.; Wang, B. Development of a monoclonal antibody-based ELISA for the detection of the novel insecticide cyantraniliprole. *RSC Adv.* **2015**, *5*, 35874–35881. [CrossRef]
23. Simon, E.; Knopp, D.; Carrasco, P.B.; Niessner, R. Development of an enzyme immunoassay for metsulfuron-methyl. *Food Agric. Immunol.* **2008**, *10*, 105–120. [CrossRef]
24. Zhao, J.; Li, G.; Wang, B.-M.; Liu, W.; Nan, T.-G.; Zhai, Z.-X.; Li, Z.-H.; Li, Q.X. Development of a monoclonal antibody-based enzyme-linked immunosorbent assay for the analysis of glycyrrhizic acid. *Anal. Bioanal. Chem.* **2006**, *386*, 1735–1740. [CrossRef] [PubMed]

25. Zhao, J.; Li, G.; Yi, G.X.; Wang, B.M.; Deng, A.X.; Nan, T.G.; Li, Z.H.; Li, Q.X. Comparison between conventional indirect competitive enzyme-linked immunosorbent assay (icELISA) and simplified icELISA for small molecules. *Anal. Chim. Acta* **2006**, *571*, 79–85. [CrossRef] [PubMed]
26. Eremin, S.; Ryabova, I.; Yakovleva, J.; Yazynina, E.; Zherdev, A.; Dzantiev, B. Development of a rapid, specific fluorescence polarization immunoassay for the herbicide chlorsulfuron. *Anal. Chim. Acta* **2002**, *468*, 229–236. [CrossRef]

Article

Electrochemical Sensor Based on Laser-Induced Graphene for Carbendazim Detection in Water

Li Wang, Mengyue Li, Bo Li, Min Wang, Hua Zhao and Fengnian Zhao *

College of Chemistry and Materials Engineering, Beijing Technology and Business University, Beijing 100048, China
* Correspondence: zhaofn@btbu.edu.cn

Abstract: Carbendazim (CBZ) abuse can lead to pesticide residues, which may threaten the environment and human health. In this paper, a portable three-electrode sensor based on laser-induced graphene (LIG) was proposed for the electrochemical detection of CBZ. Compared with the traditional preparation method of graphene, LIG is prepared by exposing the polyimide film to a laser, which is easily produced and patterned. To enhance the sensitivity, platinum nanoparticles (PtNPs) were electrodeposited on the surface of LIG. Under optimal conditions, our prepared sensor (LIG/Pt) has a good linear relationship with CBZ concentration in the range of 1–40 μM, with a low detection limit of 0.67 μM. Further, the sensor shows good recovery rates for the detection of CBZ in wastewater, which provides a fast and reliable method for real-time analysis of CBZ residues in water samples.

Keywords: carbendazim; laser-induced graphene; platinum nanoparticles; electrochemical detection; water sample

1. Introduction

Carbendazim (CBZ) is a broad-spectrum fungicide with excellent activity and durability. It has been widely used in fruits and vegetables to protect crops from pathogenic bacteria and pests [1,2]. However, as a benzimidazole cyclic compound, carbendazim has strong stability and is not easily degraded in soil for a long time, thus causing serious residue problems and endangering environmental and food safety [3]. To date, CBZ has been banned in many countries, but it is still allowed in China [4]. The Ministry of Health and the Ministry of Agriculture have set maximum residue limits (MRLs) for CBZ in food to ensure that CBZ residues will not threaten human health. Therefore, it is of great significance to establish a rapid, sensitive, and effective method for the determination of CBZ residues for food safety [5].

At present, many methods have been developed to detect CBZ, including high-performance liquid chromatography (HPLC) [6], ultraviolet–visible spectrophotometry [7], capillary electrophoresis [8], fluorescence spectrometry [9], and electrochemical analysis [10]. Although sensitive and reliable, some methods have some drawbacks, such as complex sample pretreatment, high operation requirements, long detection time, and expensive instruments which are difficult to apply to rapid detection and on-site analysis of pesticide residues [11,12]. As a supplement, the electrochemical method is widely used in the detection of pesticide residues because of its low cost, high sensitivity, and simple operation [13,14]. At present, various electrochemical sensing methods have been used for the detection of CBZ [15]. To enhance analytical performance, carbon-based materials have been widely used in the field of electrochemical analysis due to their superior electrical properties [16]. Among them, graphene is a typical carbon-based material with a large surface area and favorable conductivity, which has been regarded as an excellent electrode material [17,18]. For example, Kumar et al. [19] used the nitrogen-doped reduced graphene oxide (RGO) as the nanocarrier to anchor gadolinium sesquisulfide for the electrochemical detection of CBZ in the river water sample. Further, Sundaresan et al. [20] constructed a

Citation: Wang, L.; Li, M.; Li, B.; Wang, M.; Zhao, H.; Zhao, F. Electrochemical Sensor Based on Laser-Induced Graphene for Carbendazim Detection in Water. *Foods* **2023**, *12*, 2277. https://doi.org/10.3390/foods12122277

Academic Editor: Paola Roncada

Received: 30 April 2023
Revised: 27 May 2023
Accepted: 2 June 2023
Published: 6 June 2023

highly sensitive electrochemical sensor using RGO as the functional platform to embed tungstate nanostructure for the detection of fenitrothion. Although these electrode materials have high sensitivity and good detection performance, the preparation process of graphene electrodes is complicated. To simplify the preparation of graphene electrodes, some advanced printing techniques were developed to broaden the application of graphene-based sensors in the analytical field [21–23].

Laser-induced graphene (LIG) is a porous graphene material with high electrocatalytic activity, a large surface area, and three-dimensional morphology [24]. The patterned LIG-based electrode can be easily obtained by scanning the surface of thermoplastic polymer materials (such as polyimide, PI) with the laser [25]. Previously, some low-cost, fast, sensitive, and flexible LIG-based electrochemical sensors were prepared on a polymer substrate by a laser direct engraving process [26]. However, single LIG devices usually exhibit limited performance sensitivity [27,28]. To solve this problem, various metal materials, especially platinum, silver, and gold nanoparticles (NPs), can be embedded in carbon carriers as connecting materials to improve sensitivity [29,30]. For example, Wang et al. [31] developed a label-free carcinoembryonic antigen (CEA) electrochemical immunosensor by anchoring AuNPs to LIG (LIG/Au) using chloroauric acid as a precursor. You et al. [30] fabricated laser-induced noble metal NPs (such as AuNPs, AgNPs, and PtNPs) and graphene composite, which were further applied to obtain the flexible electrode to realize the electrochemical detection of the pathogen.

Inspired by this, a strategy of detecting CBZ by the portable and integrated three-electrode electrochemical sensor based on LIG/Pt was put forward (Figure 1). Herein, three-dimensional porous LIG electrodes were prepared by laser direct writing on flexible PI thin films. To improve the sensitivity of the detection system, a PtNP-modified LIG sensor (LIG/Pt) was prepared by the electrodeposition method on porous LIG composites. Compared with the unmodified LIG electrode, the oxidation peak of CBZ on LIG/Pt was obviously enhanced, which proved that the prepared sensor had good electrochemical analysis performance. By connecting the prepared sensor with a handheld electrochemical workstation, the real-time information on CBZ residues was received on a smartphone, which could provide a simple, portable, and sensitive method for rapid and on-site analysis of CBZ in water samples.

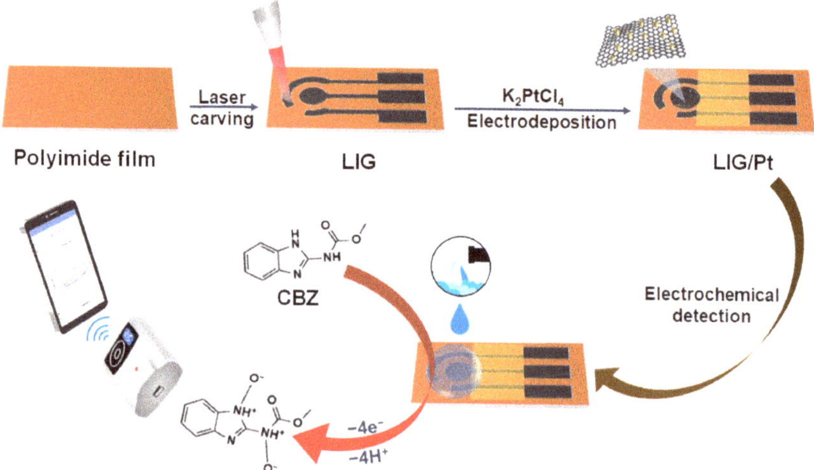

Figure 1. Preparation of LIG/Pt electrochemical sensor and its application for the real-time detection of CBZ in wastewater samples.

2. Materials and Methods

2.1. Reagents and Instruments

CBZ, monocrotophos, methyl parathion, and phosphamidon were purchased from Shanghai Aladdin Bio-Chem Technology Co., Ltd. (Shanghai, China). Potassium chloroplatinite (K_2PtC1_4) was obtained from Shanghai Macklin Biochemical Co., Ltd. (Shanghai, China). Phosphate buffer solution (PBS, 0.1 M, pH 7.4) was purchased from Beijing Bairuiji Biotechnology Co., Ltd. (Beijing, China). The commercial PI film (thickness of 125 µm) and PI tape (80 µm) were purchased from DuPont, Wilmington, DE, USA. The Ag/AgCl paste was purchased from Ercon Inc., Wareham, MA, USA. All other chemicals and reagents used were of analytical grade.

The laser etching micromachine for preparing the LIG electrode was from Tianjin Jiayin Nanotechnology Co., Ltd. (Tianjin, China). All electrochemical measurements were carried out on a hand-held electrochemical workstation (EmStat3 Blue, PalmSens BV, Houten, The Netherlands) with a wireless Bluetooth transmission module. A field emission scanning electron microscope (HITACHI SU8010, Tokyo, Japan) was used for the morphology characterization of the prepared electrodes. A LabRAM HR Evolution Raman microscope system (Horiba Jobin Yvon, Kyoto, Japan) was operated to analyze the chemical composition of the porous graphene. An ESCALAB 250 X-ray photoelectron spectroscopy (ThermoFisher SCIENTIFIC, Waltham, MA, USA) was used to analyze the surface elemental composition of porous graphene. A 1200 liquid chromatograph (Agilent, Santa Clara, CA, USA) equipped with an AB5000 mass spectrometer (SCIEX, Boston, MA, USA) was taken as the gold standard.

2.2. Preparation of the LIG/Pt Sensor

In this experiment, the LIG electrode with porous structure was prepared by laser direct writing technique (Figure 1). The specific steps are as follows. The LIG-based three-electrode pattern was generated by using a computer to control a laser system to scan the surface of the PI film (power: 1.38 W; speed: 4.0 cm s^{-1}). Then, the Ag/AgCl slurry was coated on the reference electrode and dried at 70 °C for 1 h. Finally, the electrode was encapsulated with the PI tape, thus realizing the preparation of the three-electrode LIG sensor.

The electrodeposition process of PtNPs was studied by the cyclic voltammetry (CV) method. 70 µL of 2 mM K_2PtC1_4 (solvent is 0.1 M Na_2SO_4) was dripped on the electrode area, and the parameters were set (scanning potential was from −0.4 V to +0.5 V, scanning rate was 50 mV s^{-1}) with 20 cycles to prepare the LIG/Pt sensor.

2.3. Electrochemical Detection of CBZ

In this work, the square wave voltammetry (SWV) method was used to detect CBZ at room temperature, and all experiments were carried out on the sensor of a three-electrode LIG/Pt system combined with a hand-held electrochemical workstation. The electrochemical detection of CBZ mainly involves the following steps: firstly, clean the surface of LIG/Pt electrode with distilled water and dry it, and connect it with a hand-held electrochemical workstation. Then, a certain concentration of CBZ in 0.1 M PBS solution was dropped on the working area of the electrode. Finally, the SWV curve was recorded in the potential window of +0.4 to +1.0 V with the equilibrium equipment time of 15 s and the detection frequency of 10 Hz, respectively.

2.4. Real Sample Analysis

In this study, wastewater samples from a pig slaughterhouse in Hebei province were collected and used for real sample analysis. The water sample was diluted 5 fold (i.e., 1 mL of water sample was diluted with 4 mL 0.1 M PBS). Three groups of different spiking levels (10, 20, 30 µM) of CBZ were added to the water sample. Then, the mixture was vortexed and filtered through a 0.22 µm filter membrane. The supernatant was collected and used for the electrochemical analysis.

LC–MS/MS method was adopted to evaluate the accuracy. The analytical column used in chromatography is Waters XBridge C_{18} column, with 0.1% acetonitrile (A) and 0.1% formic acid solution (B) as the mobile phases, the column temperature was 40 °C, the flow rate was 200 μL min^{-1}, and the injection volume was 10 μL, the gradient elution procedures are 80–60% B (0–2.0 min), 60–10% B (2.0–2.5 min), 10% B (2.5–3.0 min), 10–90% B (3.0–3.5 min) and 90% B (3.5–4.0 min). The mass spectrometry adopts an electrospray ionization (ESI) probe, the scanning mode was negative ion mode, and the temperature was 300 °C.

3. Results and Discussion

3.1. Characterization of the LIG Electrode

The preparation process of LIG is that a tightly focused laser beam can generate a high enough temperature on the target material to break its chemical bonds and rearrange carbon atoms into hexagonal graphene with characteristics [25,32]. In this study, the LIG-based three-electrode sensor was prepared by laser direct writing on the PI film as the substrate. During this process, the engraving power and scanning speed are the main influencing factors. Here, the engraving power of the laser is first characterized. When the power is too low (~1.1 W), the electrode pattern would not be completely printed. With the increase in laser power, the pattern on the PI film will be clearer. However, excessive power (~1.65 W) could damage the pattern and even puncture the PI film. Therefore, we first optimized the power with the same scanning speed (4.0 cm s^{-1}). Three groups of LIG prepared with the most suitable engraving power were characterized by CV (Figure 2a) and differential pulse voltammetry (DPV, Figure S1a) in 0.1 M KCl solution containing 1.0 mM $K_3[Fe(CN)_6]$. Results show that the electrochemical behavior of the LIG electrode is better at 1.38 W power. Then, the scanning speed was optimized under the optimal powder. Theoretically, the lower the scanning speed, the more graphene is formed. However, Figures 2b and S1b show that the LIG electrical activity at a scanning speed of 4.0 cm s^{-1} is higher than 2.4 cm s^{-1}, which may be due to the collapse of the electrode structure at the slower scanning speed [33]. The size parameters of the LIG electrode are shown in Figure S2a. The diameter of the working electrode area is 3.00 mm, and the length of the electrode sensing line is 5.25 mm. As shown in Figure S2b, the LIG electrode prepared under the optimized conditions has a smooth silver-gray surface, and its size and morphology are consistent with the parameters involved.

After that, the morphology of the prepared LIG under the optimized conditions was characterized by SEM. As shown in Figures 2c and S1c, the fabricated LIG electrode appears typically porous structure, which is formed by the graphene sheets during laser scanning [27]. Raman spectroscopy and X-ray photoelectron spectroscopy (XPS) were used to characterize the chemical composition of the obtained materials. As shown in Figure 2d, three prominent peaks are observed in the Raman spectrum of LIG, namely, the D peak (~1350 cm^{-1}), the G peak (~1580 cm^{-1}), and the 2D peak (~2700 cm^{-1}). Generally, the D peak reflects the lattice defects in graphene, while the G peak and the 2D peak reflect the characteristics of the prepared materials. By comparing the typical Raman spectra of graphene [34], it can be concluded that the materials we prepared belong to graphene materials. The C 1s spectrum of XPS analysis of LIG is shown in Figure S3. The main component of LIG, namely sp^2-hybridized graphite carbon (C=C) [35], is presented at 284.5 eV, and some disordered carbon (C-C) is also observed at 285.5 eV. In addition, a small number of C-O-C groups were also observed at 287.5 eV.

3.2. Characterization of the LIG/Pt Sensor

Laser direct writing technology can be used to prepare LIG with controllable surface morphology, surface properties, chemical composition, and electrical properties, which can also minimize the use of raw materials. While noble NPs have the function of signal amplification, and thus embedding precious metal NPs on the surface of the LIG electrode will make it higher sensitivity and electrical activity [36]. In this paper, PtNPs were electrodeposited on the surface of LIG to form the LIG/Pt sensor. Here, several LIG/Pt sensors were prepared under different deposition cycles by the CV method. As shown in Figures 3a and S4a, the CV and DPV responses are increased after the modification of PtNPs, compared to those of bare LIG electrodes. When the electrodeposition cycles are 20, the response of the LIG/Pt sensor (labeled as LIG/Pt-20) reaches the most. As shown in Figures 3b and S4b, PtNPs (sizes of 70 ± 16 nm) are evenly distributed on the surface of LIG, which can increase the surface area of the electrode and further enhance the sensitivity.

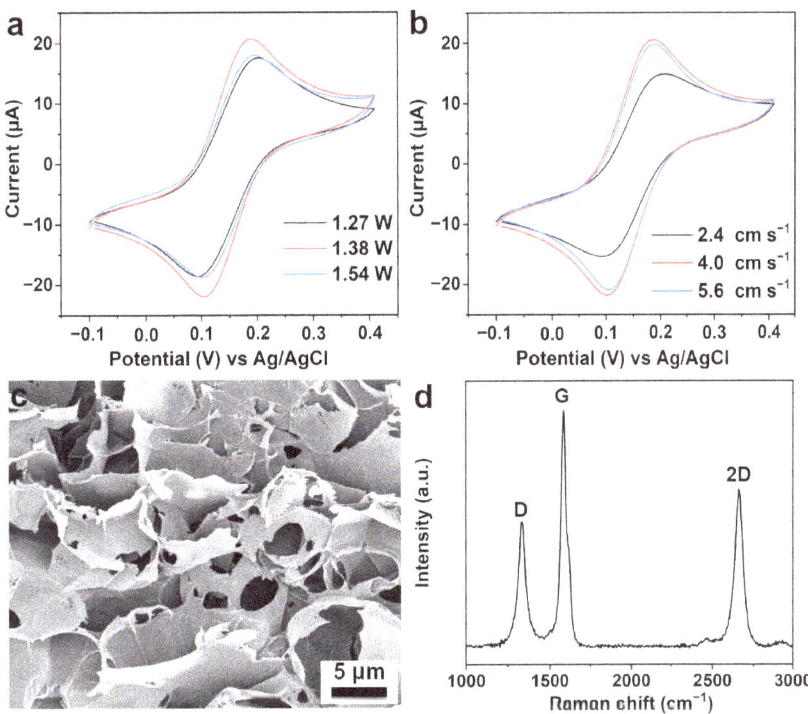

Figure 2. Characterization of the prepared LIG electrode. CV curves of the prepared electrodes with various (**a**) power and (**b**) scanning speed parameters in 0.1 M KCl solution containing 1.0 mM $K_3[Fe(CN)_6]$. (**c**) SEM image and (**d**) Raman spectrum of the prepared LIG electrode under optimal conditions (1.38 W and 4.0 cm s^{-1}).

After that, we investigated the electrochemical behavior of the obtained sensors for CBZ. Figure 3c shows the CV responses of the LIG/Pt sensor and bare LIG sensor with or without 10 μM CBZ. It is worth noticing that the LIG/Pt sensor has a higher response current than the bare electrode, which is because PtNPs can accelerate the electron transfer rate and amplify the background signal of LIG. After dropping 10 μM CBZ on the electrode surface, an obvious oxidation peak (O peak) appears at +0.77 V on the CV curve, which indicates that CBZ is oxidized to methyl carbamate and corresponding benzimidazole radical on the LIG/Pt sensor by the CV method [37]. The oxidation process of CBZ involves four electrons, and the possible oxidation mechanism is shown in Figure 3d. In contrast, the bare LIG sensor shows a low peak response due to its slow electron transfer rate. According to the above results, it can be seen that our prepared LIG/Pt sensor shows a favorable potential in the detection of CBZ.

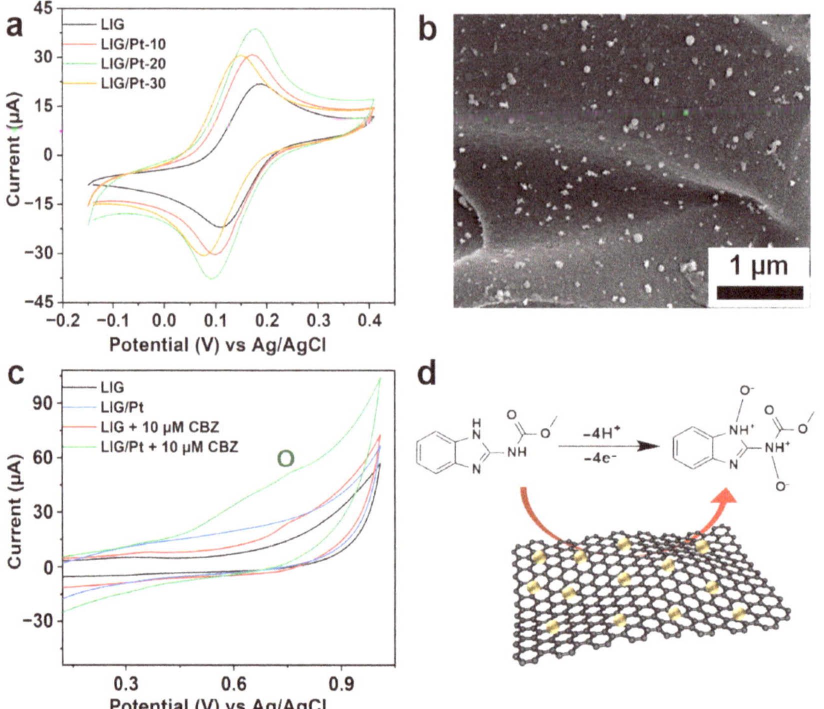

Figure 3. Characterization of the prepared LIG/Pt sensor. (**a**) Optimization of the electrodeposition cycles of the LIG/Pt sensor. (**b**) SEM image of the LIG/Pt sensor under optimal conditions. (**c**) CV responses of the LIG/Pt and bare LIG with or without 10 μM CBZ. (**d**) The electrochemical detection mechanism of CBZ.

3.3. Optimization of Experimental Parameters

Although the sensitivity of the SWV method may be lower than that of the DPV method, the faster analytical speed makes it unique in the rapid and on-site analysis. Therefore, the determination performance of the LIG/Pt sensor was studied via the SWV method in this experiment. Herein, the equilibrium time and scanning frequency were optimized to obtain the optimal instrumental parameters. Firstly, we set various equilibrium time (0 s, 10 s, 15 s, 20 s, 25 s) with the same frequency of 10 Hz. The original data and calibrated data of SWV are shown in Figure 4a,d. Results indicate that the current response of SWV for 10 μM CBZ is better when the equilibrium time is 15 s. After that, the scanning frequency was optimized with the optimal equilibrium time of 15 s. Figure 4b,e show the original data and calibrated data of SWV response signals at different frequencies. It is clear that the current response increases with the increment of scanning frequency. When the scanning frequency reaches 10 Hz, the current response increases to a maximum. Thus, we choose 10 Hz as the optimal frequency for CBZ detection.

Finally, the SWV response of CBZ was optimized in various pH values of the working solution (0.1 M PBS). As shown in Figure 4c, with the increase in pH, the peak potential shifts negatively, which indicates that protons may participate in the reaction process at the modified electrode. In addition, by observing the calibrated image (Figure 4f), it can be seen that with the increase in pH value, the peak current first increases and then decreases. Therefore, the optimal pH value for the working solution was 7.

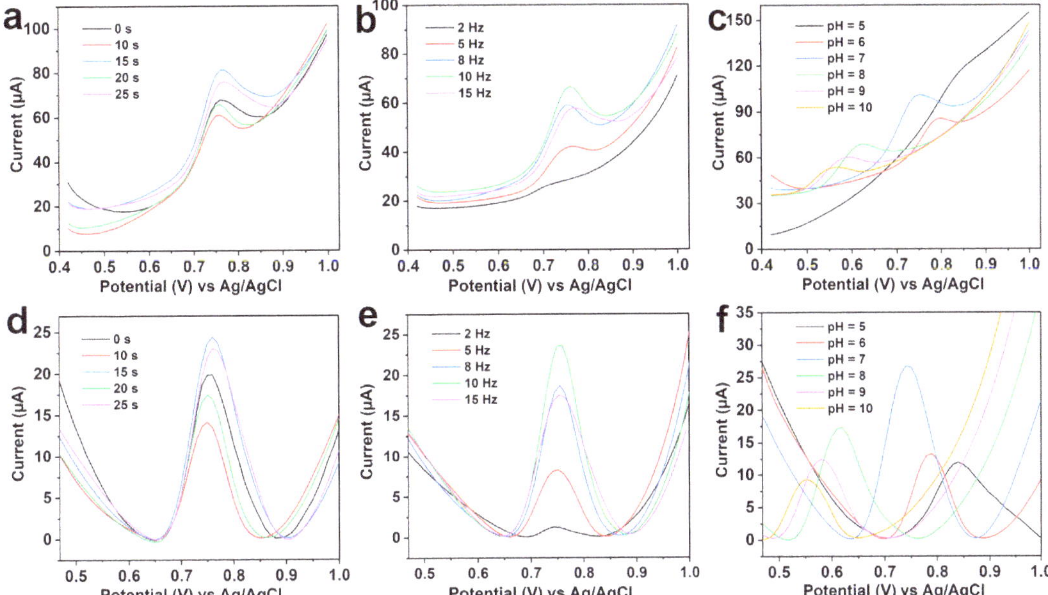

Figure 4. Parameter optimization of SWV responses of the LIG/Pt sensor for CBZ. The original SWV response data with various (**a**) equilibrium times, (**b**) detection frequencies, and (**c**) pH values of 0.1 M PBS solution. (**d–f**) The corresponding baseline-corrected data.

3.4. Electrochemical Detection of CBZ

Under the optimized conditions, the corresponding relationship between the CBZ concentration and SWV current signal was investigated. As shown in Figure 5a, the peak current response gradually increases with the increase in CBZ concentration. To obtain the current peak more conveniently, Origin software was used to correct the curves. Based on the baseline-corrected data in Figure 5b, a calibration curve is fitted with CBZ concentration as abscissa and current response as ordinate (Figure 5c). Results indicate that when the concentration of CBZ is in the range of $1-40$ μM, the corresponding current signal has a linear relationship with its concentration, the regression equation is $y = 2.0142x + 0.0022$, and the correlation coefficient R^2 is 0.9998. According to the formula of detection limit (LOD) = 3 Sb/m (Sb is the standard deviation of SWV in the blank experiment and m is the slope of the calibration curve), the LOD is 0.67 μM (i.e., 128.1 μg L^{-1}), which is far below than the MRLs stated by China in many food samples, such as some vegetables and fruits.

In addition, the performance of our sensor was also compared with other reported electrochemical sensors [11,37–39]. As shown in Table 1, our prepared LIG/Pt sensor has comparable analytical performance for the detection of CBZ.

3.5. Anti-Interference, Selectivity, Reproducibility, and Stability Analysis

In the real sample analysis, some non-target components may interfere with the detection process due to the complexity of the sample matrix. Firstly, we investigated the anti-interference ability of the LIG/Pt sensor. Several possible ions (Ca^{2+}, K$^+$, Na$^+$, 100-fold higher than CBZ) and small molecules (glucose and ascorbic acid, 50-fold higher than CBZ) that may coexist or exist during the real sample analysis were selected in this study. As shown in Figure S5a, there is no obvious difference in electrical signals in the experiment which indicates the favorable anti-interference performance of our LIG/Pt sensor.

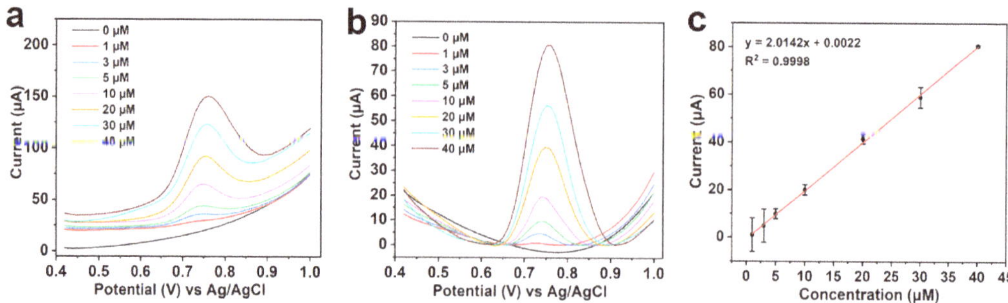

Figure 5. SWV responses of the LIG/Pt sensor for CBZ with the concentration in the range of 0–40 μM. (**a**) The SWV response data initial image. (**b**) The SWV response data with baseline correction. (**c**) The corresponding calibration curve between CBZ concentration (1–40 μM) and SWV current signal.

Table 1. Comparison of CBZ determination by electrochemical sensors with different electrodes.

Sensor	Method	Linear Range	LOD	Ref.
GCE/PPy-CNT	DPV	4–20 μM	1.3 μM	[39]
GCE/MBC@CTS	DPV	0.1–20 μM	0.02 μM	[11]
GCE/RGO-Pt	DPV	25–115 μM	2.96 μM	[38]
GCE/RGO/NP-Cu	DPV	0.5–30 μM	0.09 μM	[37]
GCE/NiCo-LDH	DPV	0.006–14.1 μM	0.001 μM	[40]
LIG/Pt	SWV	1–40 μM	0.67 μM	This work

Notes: GCE, Glassy carbon electrode; PPy, polypyrrole; CNT, carbon nanotube; MBC@CTS, mung bean-derived porous carbon@chitosan; RGO, reduced graphene oxide; NP-Cu, nanoporous copper; NiCo-LDH, nickel cobalt-layered double hydroxide.

Moreover, we evaluated the selectivity of our sensor. Herein, the SWV signals of the LIG/Pt were recorded in the same concentration (10 μM) of CBZ, monocrotophos, methyl parathion, and phosphamidon, respectively. As shown in Figure S5b, the LIG/Pt only exhibits a clear SWV signal for CBZ, indicating the high selectivity of our sensor for CBZ.

Reproducibility and stability are important performance parameters of the electrode. Five electrodes were prepared under the same conditions, and the reproducibility of the LIG/Pt electrode was evaluated by detecting 10 μM CBZ with the SWV method. As shown in Figure S5c, the current response ratio of five electrodes has little difference, with the RSD not over 1.79%, indicating the favorable repeatability of our sensor. Further, the stability of the LIG/Pt sensor was assessed by storing it for several days at room temperature. Results show that the current response ratios in five days are from 94.8% to 103.8%, with the RSD below 3.26%, declaring the satisfying storage stability of our sensor (Figure S5d).

3.6. Real Sample Analysis

In this paper, wastewater was selected as the real sample, and the recovery experiment was adopted for methodological evaluation. We set three spiking levels (10, 20, 30 μM) in blank water samples. After placing for 0.5 h at room temperature, the above samples were diluted 5 fold with 0.1 M PBS (pH = 7) and then filtered through a 0.22 μm membrane. The supernatant was collected for detection. As shown in Table 2, the recoveries of CBZ in wastewater range from 88.89% to 99.50%, with RSDs of 1.98–5.11% (*n* = 3). To evaluate the accuracy of the sensor detection, the CBZ concentration was measured simultaneously by the LC–MS/MS standard method. The results show that the determination of CBZ in water samples by LIG/Pt is similar to that by LC–MS/MS, indicating the satisfying accuracy of our prepared LIG/Pt sensor. In the future, our efforts will be directed towards applying this sensor in more complicated food samples by the improved sample pretreatment method, such as the adoption of efficient cleaning agents.

Table 2. Recovery study of CBZ in wastewater samples ($n = 3$).

Method	Added (µM)	Founded (µM)	Recovery (%)	RSD (%)
	10	9.95 ± 0.20	99.50	1.98
This method	20	17.78 ± 1.61	88.92	4.12
	30	26.67 ± 1.10	88.89	5.11
	10	8.51 ± 0.11	85.14	1.27
LC–MS/MS	20	17.44 ± 0.55	87.22	3.14
	30	27.07 ± 1.69	90.24	6.23

4. Conclusions

In this study, a portable LIG-based electrochemical sensor was developed for the rapid detection of CBZ in water samples. By laser direct writing on the PI film, the LIG three-electrode sensor was easily obtained. The structure and composition of LIG were verified by SEM, Raman spectroscopy, and XPS, which confirmed its highly porous graphene structure and excellent specific surface area. By electrodepositing PtNPs on the surface of LIG, the detection performance was further enhanced. Under optimal conditions, the prepared sensor (LIG/Pt) has a wide linear range (1–40 µM), a satisfactory LOD (0.67 µM), and good recoveries (88.89–99.50%) in wastewater samples. The electrochemical sensor prepared in this study is simple in operation, high in sensitivity, and good in selectivity, which can provide a reliable and real-time analysis method for the detection of CBZ residues in water samples.

Supplementary Materials: The following supporting information can be downloaded at: https://www.mdpi.com/article/10.3390/foods12122277/s1, Figure S1: Characterization of the prepared LIG electrode. DPV curves of the LIG electrode prepared by optimizing (a) power and (b) scanning speed. (c) SEM image of porous LIG; Figure S2: Images of the LIG-based three-electrode sensor. (a) The size parameters and (b) the physical image of the LIG electrode; Figure S3: XPS fittings for C 1s of LIG; Figure S4: Characterization of the prepared LIG/Pt sensor. (a) DPV response of LIG/Pt sensors with different electrodeposition cycles (10, 20, and 30 are the CV cycles during the electrodeposition). (b) SEM image of the prepared LIG/Pt sensor; Figure S5: Measurement of (a) anti-interference performance, (b) selectivity, (c) reproducibility, and (d) stability of the LIG/Pt sensor.

Author Contributions: L.W., methodology, validation, formal analysis, investigation, and writing—original draft. M.L., methodology, validation, and formal analysis. B.L., methodology, validation, and formal analysis. M.W. and H.Z., supervision. F.Z., conceptualization, supervision, funding acquisition, project administration, and writing—review and editing. All authors have read and agreed to the published version of the manuscript.

Funding: This research was supported by the National Natural Science Foundation of China (grant number 32202160) and the Research Foundation for Youth Scholars of Beijing Technology and Business University (grant number QNJJ2022-27).

Data Availability Statement: The data used to support the findings of this study can be made available by the corresponding author upon request.

Conflicts of Interest: The authors declare no conflict of interest.

References

1. Haipeng, G.; Ruiqi, Y.; Doudou, L.; Xiao, M.; Dingkun, Z.; Peng, L.; Jiabo, W.; Lidong, Z.; Weijun, K. Red-emissive carbon dots based fluorescent and smartphone-integrated paper sensors for sensitive detection of carbendazim. *Microchem. J.* **2023**, *190*, 108586.
2. Martins, T.S.; Machado, S.A.S.; Oliveira, O.N., Jr.; Bott-Neto, J.L. Optimized paper-based electrochemical sensors treated in acidic media to detect carbendazim on the skin of apple and cabbage. *Food Chem.* **2023**, *410*, 135429. [CrossRef] [PubMed]
3. Silva, A.R.R.; Cardoso, D.N.; Cruz, A.; Mendo, S.; Soares, A.; Loureiro, S. Long-term exposure of Daphnia magna to carbendazim: How it affects toxicity to another chemical or mixture. *Environ. Sci. Pollut. Res.* **2019**, *26*, 16289–16302. [CrossRef] [PubMed]
4. Yamuna, A.; Chen, T.W.; Chen, S.M. Synthesis and characterizations of iron antimony oxide nanoparticles and its applications in electrochemical detection of carbendazim in apple juice and paddy water samples. *Food Chem.* **2022**, *373 Pt B*, 131569. [CrossRef]

5. Fu, R.; Zhou, J.; Liu, Y.; Wang, Y.; Liu, H.; Pang, J.; Cui, Y.; Zhao, Q.; Wang, C.; Li, Z.; et al. Portable and quantitative detection of carbendazim based on the readout of a thermometer. *Food Chem.* **2021**, *351*, 129792. [CrossRef]
6. Scheel, G.L.; Teixeira Tarley, C.R. Simultaneous microextraction of carbendazim, fipronil and picoxystrobin in naturally and artificial occurring water bodies by water-induced supramolecular solvent and determination by HPLC-DAD. *J. Mol. Liq.* **2020**, *297*, 111897. [CrossRef]
7. Passos, M.L.C.; Saraiva, M.F.S.M.L. Detection in UV-visible spectrophotometry: Detectors, detection systems, and detection strategies. *Measurement* **2019**, *135*, 896–904. [CrossRef]
8. Bernardo-Bermejo, S.; Sánchez-López, E.; Castro-Puyana, M.; Marina, M.L. Chiral Capillary Electrophoresis. *Trends Anal. Chem.* **2020**, *124*, 115807. [CrossRef]
9. Chen, H.; Hu, O.; Fan, Y.; Xu, L.; Zhang, L.; Lan, W.; Hu, Y.; Xie, X.; Ma, L.; She, Y.; et al. Fluorescence paper-based sensor for visual detection of carbamate pesticides in food based on CdTe quantum dot and nano ZnTPyP. *Food Chem.* **2020**, *327*, 127075. [CrossRef]
10. Zhong, W.; Gao, F.; Zou, J.; Liu, S.; Li, M.; Gao, Y.; Yu, Y.; Wang, X.; Lu, L. MXene@Ag-based ratiometric electrochemical sensing strategy for effective detection of carbendazim in vegetable samples. *Food Chem.* **2021**, *360*, 130006. [CrossRef]
11. Liu, R.; Chang, Y.; Li, F.; Dubovyk, V.; Li, D.; Ran, Q.; Zhao, H. Highly sensitive detection of carbendazim in juices based on mung bean-derived porous carbon@chitosan composite modified electrochemical sensor. *Food Chem.* **2022**, *392*, 133301. [CrossRef] [PubMed]
12. Wang, Z.; Li, S.; Hu, P.; Dai, R.; Wu, B.; Yang, L.; Huang, Y.; Zhuang, G. Recent developments in the spectrometry of fluorescence, ultraviolet visible and surface-enhanced Raman scattering for pesticide residue detection. *Bull. Mater. Sci.* **2022**, *45*, 202. [CrossRef]
13. Rashmi Dilip, K.; Rangappa, S.K.; Nagaraju, D.H.; Srinivasa, B. State-of-the-art Electrochemical Sensors for Quantitative Detection of Pesticides. *Appl. Organomet. Chem.* **2023**, *37*, e7097.
14. Zheng, X.; Khaoulani, S.; Ktari, N.; Lo, M.; Khalil, A.M.; Zerrouki, C.; Fourati, N.; Chehimi, M.M. Towards Clean and Safe Water: A Review on the Emerging Role of Imprinted Polymer-Based Electrochemical Sensors. *Sensors* **2021**, *21*, 4300. [CrossRef]
15. Suresh, I.; Selvaraj, S.; Nesakumar, N.; Rayappan, J.B.B.; Kulandaiswamy, A.J. Nanomaterials based non-enzymatic electrochemical and optical sensors for the detection of carbendazim: A review. *Trends Environ. Anal. Chem.* **2021**, *31*, e00137. [CrossRef]
16. Anil Kumar, Y.; Koyyada, G.; Ramachandran, T.; Kim, J.H.; Sajid, S.; Moniruzzaman, M.; Alzahmi, S.; Obaidat, I.M. Carbon Materials as a Conductive Skeleton for Supercapacitor Electrode Applications: A Review. *Nanomaterials* **2023**, *13*, 1049. [CrossRef]
17. Nithya, V.D. A review on holey graphene electrode for supercapacitor. *J. Energy Storage* **2021**, *44*, 103380. [CrossRef]
18. Aiswaria, P.; Mohamed, S.N.; Singaravelu, D.L.; Brindhadevi, K.; Pugazhendhi, A. A review on graphene / graphene oxide supported electrodes for microbial fuel cell applications: Challenges and prospects. *Chemosphere* **2022**, *296*, 133983.
19. Yogesh Kumar, K.; Prashanth, M.K.; Parashuram, L.; Palanivel, B.; Alharti, F.A.; Jeon, B.H.; Raghu, M.S. Gadolinium sesquisulfide anchored N-doped reduced graphene oxide for sensitive detection and degradation of carbendazim. *Chemosphere* **2022**, *296*, 134030. [CrossRef]
20. Sundaresan, R.; Mariyappan, V.; Chen, T.-W.; Chen, S.-M.; Akilarasan, M.; Liu, X.; Yu, J. One-dimensional rare-earth tungstate nanostructure encapsulated reduced graphene oxide electrocatalyst-based electrochemical sensor for the detection of organophosphorus pesticide. *J. Nanostructure Chem.* **2023**. [CrossRef]
21. Silva, L.R.G.; Stefano, J.S.; Orzari, L.O.; Brazaca, L.C.; Carrilho, E.; Marcolino-Junior, L.H.; Bergamini, M.F.; Munoz, R.A.A.; Janegitz, B.C. Electrochemical Biosensor for SARS-CoV-2 cDNA Detection Using AuPs-Modified 3D-Printed Graphene Electrodes. *Biosensors* **2022**, *12*, 622. [CrossRef] [PubMed]
22. Carey, T.; Alhourani, A.; Tian, R.Y.; Seyedin, S.; Arbab, A.; Maughan, J.; Siller, L.; Horvath, D.; Kelly, A.; Kaur, H.; et al. Cyclic production of biocompatible few-layer graphene ink with in-line shear-mixing for inkjet-printed electrodes and Li-ion energy storage. *Npj 2d Mater. Appl.* **2022**, *6*, 3. [CrossRef]
23. Gao, Y.; Ding, J. 3D Printed Thick Reduced Graphene Oxide: Manganese Oxide/Carbon Nanotube Hybrid Electrode with Highly Ordered Microstructures for Supercapacitors. *Adv. Mater. Technol.* **2022**, *8*, 2200263. [CrossRef]
24. Qiwen, Z.; Fangyi, Z.; Xing, L.; Zengji, Y.; Xi, C.; Zhengfen, W. Doping of Laser-Induced Graphene and Its Applications. *Adv. Mater. Technol.* **2023**, 2300244. [CrossRef]
25. Ye, R.; James, D.K.; Tour, J.M. Laser-Induced Graphene. *Acc. Chem. Res.* **2018**, *51*, 1609–1620. [CrossRef]
26. Settu, K.; Chiu, P.T.; Huang, Y.M. Laser-Induced Graphene-Based Enzymatic Biosensor for Glucose Detection. *Polymers* **2021**, *13*, 2795. [CrossRef]
27. Liu, W.; Chen, Q.; Huang, Y.; Wang, D.; Li, L.; Liu, Z. In situ laser synthesis of Pt nanoparticles embedded in graphene films for wearable strain sensors with ultra-high sensitivity and stability. *Carbon* **2022**, *190*, 245–254. [CrossRef]
28. Zhang, C.; Ping, J.; Ying, Y. Evaluation of trans-resveratrol level in grape wine using laser-induced porous graphene-based electrochemical sensor. *Sci. Total Environ.* **2020**, *714*, 136687. [CrossRef]
29. Hui, X.; Xuan, X.; Kim, J.; Park, J.Y. A highly flexible and selective dopamine sensor based on Pt-Au nanoparticle-modified laser-induced graphene. *Electrochim Acta* **2019**, *328*, 108586. [CrossRef]
30. You, Z.; Qiu, Q.; Chen, H.; Feng, Y.; Wang, X.; Wang, Y.; Ying, Y. Laser-induced noble metal nanoparticle-graphene composites enabled flexible biosensor for pathogen detection. *Biosens. Bioelectron.* **2020**, *150*, 111896. [CrossRef]
31. Wang, G.; Chen, J.; Huang, L.; Chen, Y.; Li, Y. A laser-induced graphene electrochemical immunosensor for label-free CEA monitoring in serum. *Analyst* **2021**, *146*, 6631–6642. [CrossRef] [PubMed]

32. Le, T.S.D.; Phan, H.P.; Kwon, S.; Park, S.; Jung, Y.; Min, J.; Chun, B.J.; Yoon, H.; Ko, S.H.; Kim, S.W.; et al. Recent Advances in Laser-Induced Graphene: Mechanism, Fabrication, Properties, and Applications in Flexible Electronics. *Adv. Funct. Mater.* **2022**, *32*, 2205158. [CrossRef]

33. Zhao, F.; He, J.; Li, X.; Bai, Y.; Ying, Y.; Ping, J. Smart plant-wearable biosensor for in-situ pesticide analysis. *Biosens. Bioelectron.* **2020**, *170*, 112636. [CrossRef]

34. Chyan, Y.; Ye, R.; Li, Y.; Singh, S.P.; Arnusch, C.J.; Tour, J.M. Laser-Induced Graphene by Multiple Lasing: Toward Electronics on Cloth, Paper, and Food. *ACS Nano* **2018**, *12*, 2176–2183. [CrossRef]

35. Sain, S.; Roy, S.; Mathur, A.; Rajesh, V.M.; Banerjee, D.; Sarkar, B.; Roy, S.S. Electrochemical Sensors Based on Flexible Laser-Induced Graphene for the Detection of Paraquat in Water. *ACS Appl. Nano Mater.* **2022**, *5*, 17516–17525. [CrossRef]

36. Geng, H.; Vilms Pedersen, S.; Ma, Y.; Haghighi, T.; Dai, H.; Howes, P.D.; Stevens, M.M. Noble Metal Nanoparticle Biosensors: From Fundamental Studies toward Point-of-Care Diagnostics. *Acc. Chem. Res.* **2022**, *55*, 593–604. [CrossRef]

37. Tian, C.H.; Zhang, S.F.; Wang, H.B.; Chen, C.; Han, Z.D.; Chen, M.L.; Zhu, Y.Y.; Cui, R.J.; Zhang, G.H. Three-dimensional nanoporous copper and reduced graphene oxide composites as enhanced sensing platform for electrochemical detection of carbendazim. *J. Electroanal. Chem.* **2019**, *847*, 113242. [CrossRef]

38. Pham, T.S.H.; Hasegawa, S.; Mahon, P.; Guerin, K.; Dubois, M.; Yu, A.M. Graphene Nanocomposites Based Electrochemical Sensing Platform for Simultaneous Detection of Multi-drugs. *Electroanalysis* **2022**, *34*, 435–444. [CrossRef]

39. Özdokur, K.V.; Kuşcu, C.; Ertaş, F.N. Ultrasound Assisted Electrochemical Deposition of Polypyrrole-Carbon Nanotube Composite Film: Preparation, Characterization and Application to the Determination of Droxidopa. *Curr. Anal. Chem.* **2020**, *16*, 421–427. [CrossRef]

40. Kokulnathan, T.; Wang, T.-J.; Ahmed, F.; Arshi, N. Fabrication of flower-like nickel cobalt-layered double hydroxide for electrochemical detection of carbendazim. *Surf. Interfaces* **2023**, *36*, 102570. [CrossRef]

 *foods*

Article

Dissipation, Metabolism, Accumulation, Processing and Risk Assessment of Fluopyram and Trifloxystrobin in Cucumbers and Cowpeas from Cultivation to Consumption

Kai Cui [1], Shuai Guan [1], Jingyun Liang [1], Liping Fang [1], Ruiyan Ding [1], Jian Wang [1], Teng Li [1], Zhan Dong [1,*], Xiaohu Wu [2] and Yongquan Zheng [2]

[1] Institute of Quality Standard and Testing Technology for Agro-Products, Shandong Academy of Agricultural Sciences, Shandong Provincial Key Laboratory of Test Technology on Food Quality and Safety, Jinan 250100, China
[2] Institute of Plant Protection, Chinese Academy of Agricultural Sciences, Beijing 100193, China
* Correspondence: zhandongsaas@163.com

Abstract: Fluopyram and trifloxystrobin are widely used for controlling various plant diseases in cucumbers and cowpeas. However, data on residue behaviors in plant cultivation and food processing are currently lacking. Our results showed that cowpeas had higher fluopyram and trifloxystrobin residues (16.48–247.65 µg/kg) than cucumbers (877.37–3576.15 µg/kg). Moreover, fluopyram and trifloxystrobin dissipated faster in cucumbers (half-life range, 2.60–10.66 d) than in cowpeas (10.83–22.36 d). Fluopyram and trifloxystrobin were the main compounds found in field samples, and their metabolites, fluopyram benzamide and trifloxystrobin acid, fluctuated at low residue levels ($\leq$76.17 µg/kg). Repeated spraying resulted in the accumulation of fluopyram, trifloxystrobin, fluopyram benzamide and trifloxystrobin acid in cucumbers and cowpeas. Peeling, washing, stir-frying, boiling and pickling were able to partially or substantially remove fluopyram and trifloxystrobin residues from raw cucumbers and cowpeas (processing factor range, 0.12–0.97); on the contrary, trifloxystrobin acid residues appeared to be concentrated in pickled cucumbers and cowpeas (processing factor range, 1.35–5.41). Chronic and acute risk assessments suggest that the levels of fluopyram and trifloxystrobin in cucumbers and cowpeas were within a safe range based on the field residue data of the present study. The potential hazards of fluopyram and trifloxystrobin should be continuously assessed for their high residue concentrations and potential accumulation effects.

Keywords: fluopyram and trifloxystrobin; residue behavior; residue accumulation; food processing; health risk assessment

Citation: Cui, K.; Guan, S.; Liang, J.; Fang, L.; Ding, R.; Wang, J.; Li, T.; Dong, Z.; Wu, X.; Zheng, Y. Dissipation, Metabolism, Accumulation, Processing and Risk Assessment of Fluopyram and Trifloxystrobin in Cucumbers and Cowpeas from Cultivation to Consumption. *Foods* **2023**, *12*, 2082. https://doi.org/10.3390/foods12102082

Academic Editor: Paola Roncada

Received: 16 March 2023
Revised: 25 April 2023
Accepted: 19 May 2023
Published: 22 May 2023

1. Introduction

Greenhouse cultivation has gradually developed into the dominant production approach to achieve a year-round supply of various vegetables, including cucumbers and cowpeas. However, these vegetables suffer from serious plant diseases due to their semi-enclosed and comfortable environments [1], resulting in the frequent application of fungicides. Fluopyram, N-{2-[3-chloro-5-(trifluoromethyl)pyridin-2-yl]ethyl}-2-(trifluoromethyl)benzamide, (FLU, Figure 1), a new pyridyl-benzamide fungicide, acts on the enzyme succinate dehydrogenase to inhibit spore germination, mycelium growth and sporulation. Trifloxystrobin, methyl (2E)-(methoxyimino)(2-{[({(1E)-1-[3-(trifluoromethyl)phenyl]ethylidene}amino)oxy]methyl}phenyl)acetate, (TRI, Figure 1), a strobilurin fungicide, acts as a quinone outside inhibitor to inhibit the mitochondrial respiration of pathogens. FLU and TRI are often simultaneously or alternately used to control a variety of plant diseases, such as cucumber and cowpea anthracnose, cucumber powdery mildew and target spot, achieving a high control efficiency due to their strong synergistic effects [2]. However, increasing evidence confirms their potential toxic effects on mammals, such as the induction of thyroid and liver tumor formation by FLU [3,4] and the triggering of neurotoxicity and skin toxicity by TRI [5,6]. Considering the potential health hazards of FLU and TRI, the residue fate and dietary health risks associated with exposure to FLU and TRI should be studied in agricultural products following their application.

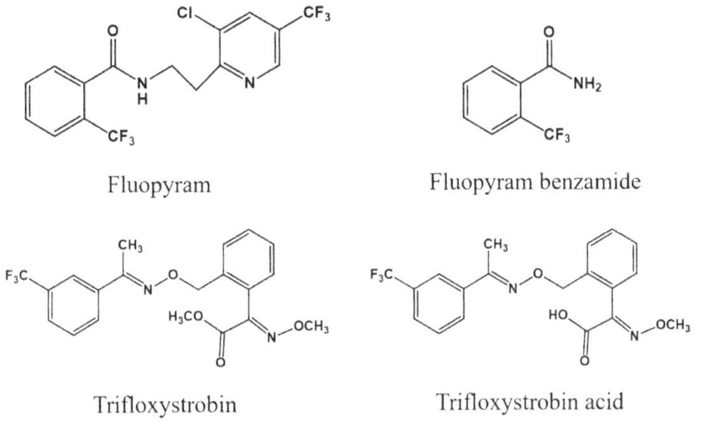

Fluopyram

Fluopyram benzamide

Trifloxystrobin

Trifloxystrobin acid

Figure 1. The chemical structures of FLU and TRI and their metabolites, FLB and TRA.

Some studies have reported on the residue levels, dissipation behavior and dietary risk assessment of FLU or TRI in different farm crops [7–13]. However, no work to date has focused on the potential accumulation of pesticide residues in crops after repeated spraying. It is generally recommended that fungicides are sprayed two to three times during crops' development periods; more alarmingly, farmers often increase their application according to disease severity and their farming experience. This may result in residue accumulation of pesticides in crops and may pose a high dietary risk to humans. Single and repeated applications of fungicides have been proven to generate different pesticide dissipation rates in crops [14]. Hence, understanding the residue dissipation and accumulation of FLU and TRI after repeated spraying is crucial to ensure food safety and protecting public health.

Agricultural products are generally processed before being eaten through operations such as peeling, washing, boiling, stir-frying and pickling. These processing operations play an important role in the reduction or increase in pesticide residues in processed agricultural products [15–19]. To date, only a few papers have been published on the changes in FLU and TRI residues during food processing. Słowik-Borowiec and Szpyrka found that washing, peeling, juicing, boiling and ultrasonic washing could remove large amounts (≥56%) of FLU from apples [20]. On the contrary, TRI residues increased 3.7–5.4 times in dried red peppers following hot-air and sunlight drying [21]. Studies have shown that parent pesticides may be metabolized into more toxic compounds during food processing [22–25]. Owing to the wide use of FLU and TRI on cucumbers and cowpeas, their main toxic metabolites, i.e., fluopyram benzamide (FLB) and trifloxystrobin acid (TRA) (Figure 1), should also be measured and assessed to achieve a comprehensive risk assessment.

Our study is the first to systematically study the dissipation, metabolism, accumulation, processing and risk assessment of FLU and TRI and their main metabolites (FLB and TRA) in cucumbers and cowpeas from cultivation to consumption. Firstly, greenhouse-field trials were conducted to investigate the dissipation behavior and accumulation potential of pesticides applied to cucumbers and cowpeas. Secondly, a preliminary study of changes in pesticide residues in cucumbers and cowpeas was conducted via several traditional household processing operations. Thirdly, the health risks associated with exposure to pesticides were assessed on the basis of our residue data. The findings of this study could help in understanding the fate of FLU and TRI residues in different scenarios and could provide a reference for the rational use of these pesticides.

2. Materials and Methods

2.1. Chemicals and Materials

Standards of FLU, FLB, TRI and TRA (all 1000 mg/L) were supplied by the Alta Scientific Co., Ltd. (Tianjin, China). Acetonitrile and formic acid used were of high-performance liquid chromatography (HPLC) grade and were acquired from Macklin Biochemical Co., Ltd. (Shanghai, China). The anhydrous magnesium sulfate ($MgSO_4$) and sodium chloride (NaCl) used were of analytical grade and were acquired from Sinopharm Chemical Reagent (Shanghai, China). The sorbents (primary, secondary amine, PSA; graphitized carbon black, GCB) and 0.22 μm nylon syringe filters were purchased from Agela Technologies (Beijing, China). All standard solutions were stored at −20 °C in the dark.

2.2. Greenhouse-Field Trials

The greenhouse-field trials were conducted at the experimental base of Shandong Academy of Agricultural Sciences (Jiyang District, Shandong, China) at 36°98′ N and 116°98′ E. No FLU or TRI had been applied to any of the experimental plots for 2 years. Each vegetable variety underwent two treatments, i.e., each had an experiment plot (100 m^2) and a control plot (50 m^2). A 43% suspension concentrate of FLU (21.5%) and TRI (21.5%) was sprayed three times every 7 days on cucumbers and cowpeas at the recommended dosage of 375 mL/ha. Three independent fresh samples (~2.0 kg) were randomly collected from each plot at 2 h, 1 d, 3 d, 5 d and 7 d after each application to study the dissipation and accumulation of pesticides. To study the effect of processing on pesticide residues, ≥20.0 kg of samples were collected 1 day after the last application. All processed and unprocessed samples were chopped, homogenized and frozen at −20 °C before analysis.

2.3. Processing Operations

The household processing procedure mimicked traditional Chinese cooking. The processing steps were conducted as follows based on modified versions of Huan et al. and An et al. [26,27].

Peeling: 1.0 kg of cucumber samples were peeled. All skins and pulps were stored separately.

Washing: 1.5 kg of samples were immersed in 4.5 L of tap water (25 °C) and then manually washed. Five sampling time points were set: 1 min, 3 min, 5 min, 7 min and 10 min.

Stir-frying: The samples were first chopped into 3–4 cm pieces. Then, 50 mL of peanut oil and 1.5 kg of chopped samples were sequentially added into an electric frying pan. The samples were stir-fried evenly at 1000 W. Five sampling time points were set: 1 min, 3 min, 5 min, 7 min and 10 min.

Boiling: 1.5 kg of chopped samples were immersed in 4.5 L of boiling tap water. Five sampling time points were set: 1 min, 3 min, 5 min, 7 min and 10 min.

Pickling: 1.5 kg of chopped samples were added to a 3.0 L sealed plastic jar, and 2.0 L of 7% NaCl aqueous solutions was then added. All jars were finally stored in an incubator at 25 °C. Six sampling time points were set: 2 h, 1 d, 3 d, 5 d, 7 d and 14 d.

After sampling, the filter papers were used to remove the excess liquids (water or oil). Finally, each sample was homogenized and frozen at −20 °C until analysis.

2.4. Analytical Procedure

The homogenized samples (10 ± 0.01 g) were weighed and added to 50 mL polypropylene centrifuge tubes. A total of 10 mL of acetonitrile (acidified with 1% formic acid, *v/v*) was added into each tube and mixed thoroughly for 5 min using a multi-tube vortex mixer (2500 r/min). To all tubes, we added 1.5 g of NaCl and 4 g anhydrous $MgSO_4$, and the tubes were shaken for an additional 1 min. After centrifuging for 5 min at 5000 r/min, the aliquot of the extract (1.5 mL) was transferred into a 2 mL cleanup tube with the addition of 20 mg GCB, 50 mg PSA and 150 mg anhydrous $MgSO_4$. The mixture was then mixed thoroughly for 1 min and centrifuged for 5 min at 5000 r/min. After filtration via a 0.22 μm nylon syringe, the resulting filtrates were prepared for subsequent HPLC-triple-quadrupole mass spectrometer (MS/MS) analysis.

For quantitative analysis of FLU and TRI and their main metabolites (FLB and TRA), a 1290 Infinity II HPLC equipped with a 6495 MS/MS (Agilent, Santa Clara, CA, USA) with a Poroshell 120 EC-C18 column (2.1 mm $\times$ 50 mm, i.d., 2.7 μm, Agilent, USA) was used. The HPLC–MS/MS was operated in electrospray-positive ionization (ESI+) and multiple reaction monitoring (MRM) modes. The column temperature and injected volume were set to 40 °C and 2 μL, respectively. The mobile phase was composed of acetonitrile (A) and 0.1% formic acid aqueous solution (B) (0.3 mL/min), with the following parameters: 0 min, 10% A; 0.5 min, 10% A; 2.5 min, 90% A; 3.5 min, 90% A; 3.6 min, 10% A; 5 min, 10% A. The ESI parameters were as follows: capillary voltage, 3.5 kV; nozzle voltage, 500 V; gas temperature, 200 °C; gas flow, 11 L/min; nebulizer, 25 psi; sheath gas temperature, 300 °C; sheath gas flow, 12 L/min. Details of the MRM conditions for the analysis of the four analytes are shown in Table S1.

2.5. Health Risk Assessment

The health risks to consumers associated with exposure to FLU and TRI were assessed based on the risk quotient (RQ) method. The national estimated daily intake (NEDI) and chronic risk quotient (RQc) were calculated to assess chronic dietary intake risk using Equations (1) and (2). Moreover, the international estimation of short-term intake (IESTI) and the acute risk quotient (RQa) were calculated to assess acute dietary intake risk using Equations (3)–(5) [28].

$$NEDI = STMR \times F/bw/1000 \tag{1}$$

$$RQc = NEDI/ADI \tag{2}$$

$$IESTI = LP \times HR/bw/1000 \tag{3}$$

$$IESTI = LP \times HR \times v/bw/1000 \tag{4}$$

$$RQa = IESTI/ARfD \tag{5}$$

In the above equations, STMR and HR represent the supervised trial median residue and the highest residue in greenhouse cowpeas and cucumbers (μg/kg), respectively. F refers to the daily intake of cowpeas and cucumbers (g/d), and bw represents the average body weight of a Chinese child or adult (kg). ADI and ARfD represent the acceptable daily intake and acute reference dose for FLU and TRI (μg/kg bw/d), respectively. LP is the large portion for cowpeas and cucumbers (g/d), and v is the variability factor. When an RQ value is larger than 1, it is considered to pose an unacceptable health risk. A higher RQ value represents a greater health risk. Notably, Equation (3) is used to calculate the IESTI for cowpeas, and Equation (4) is used for cucumbers. All parameters used for the calculation are supplied in Table S2.

2.6. Analysis of Results

The dissipation kinetics for FLU and TRI in cucumbers and cowpeas were calculated using Equation (6), and the half-life ($t_{1/2}$) was calculated using Equation (7) [29].

$$C_t = C_0 \times e^{-kt} \tag{6}$$

$$t_{1/2} = (\ln 2)/k \tag{7}$$

In Equations (6) and (7), C_t represents the sample residue at time t (μg/kg), C_0 represents the initial sample residue (μg/kg), and k refers to the dissipation rate constant.

FLU and TRI residue accumulation (RA) in cucumbers and cowpeas after repeated spraying was calculated using Equation (8) [14].

$$RA = 1: (C_2/C_1): (C_3/C_2): \ldots \ldots (C_n/C_{n-1}) \tag{8}$$

In Equation (8), C_n represents the mean residue for a pesticide at each same time point after n instances of repeated spraying ($\mu g/kg$).

The processing factor (PF) for FLU and TRI during household processing operations was calculated using Equation (9) [30].

$$PF = C_{pp}/C_{rap} \tag{9}$$

In Equation (9), C_{pp} represents the pesticide residue concentrations in the processed products ($\mu g/kg$), and C_{rap} represents the pesticide residue concentrations in the raw agricultural products ($\mu g/kg$). PF > 1 represents pesticide residue concentration and PF < 1 represents pesticide residue reduction.

The matrix effect (ME) reflects the effects of co-extractives on the signal increase/decrease in pesticides and was calculated using Equation (10) [31].

$$ME = slope\ (matrix)/slope\ (solvent) \times 100\% \tag{10}$$

In Equation (10), slope (matrix) represents the slope of the matrix calibration curve, and slope (solvent) represents the slope of the solvent calibration curve. In general, an $|ME| \leq 10\%$ represents an acceptable matrix effect. An $ME < -10\%$ represents a signal suppression effect, and an $ME > +10\%$ represents a signal enhancement effect.

The residue definition of FLU was defined as the sum of FLU and its main metabolite, FLB; it is expressed as FLU_{sum} and was calculated using Equation (11). The TRI residue was defined as the sum of TRI and its main metabolite, TRA (expressed as TRI_{sum}) and calculated using Equation (12).

$$FLU_{sum} = C_{FLU} + 1.04 \times C_{FLB} \tag{11}$$

$$TRI_{sum} = C_{TRI} + 2.10 \times C_{TRA} \tag{12}$$

In Equations (11) and (12), C_{FLU}, C_{FLB}, C_{TRI} and C_{TRA} ($\mu g/kg$) represent the residue concentrations of FLU, FLB, TRI and TRA. In addition, a 1/2 limit of quantification (LOQ) value is assigned to the residue concentrations that are below the LOQs.

3. Results

3.1. Method Validation

In our study, the methods for the simultaneous analysis of FLU and TRI and their metabolites, FLB and TRA, were validated in terms of linearity, LOQ, ME, precision and accuracy. Linearity was estimated based on the solvent and matrix standard calibration curves in the range of 1 $\mu g/L$ to 5000 mg/L, with satisfactory correlation coefficients (R^2) ranging from 0.9943 to 1. The LOQs, defined as the lowest spiked concentrations in the matrix samples, were 1 $\mu g/kg$ for the four analytes in the raw and processed cucumbers and cowpeas (Table S3). The ME values ranged from -79.41% to 16.00% for different matrices (Table S3), and therefore, matrix standard matrix curves were used to determine the four analytes and eliminate the influence of matrix effects. The precision and accuracy of this method were estimated based on a recovery assay experiment, in which we assessed the terms of the recoveries and the relative standard deviation (RSD). The average recoveries of the four analytes at the four spiked concentrations ranged from 83.73% to 115.78% in different matrices, with RSDs of 0.48–14.38% (Figure 2) (validation criteria: recovery of 70–120%, RSD $\leq$ 20%). These results confirm that the proposed analytical method was suitable for the extraction and detection of FLU and TRI and their two metabolites in cucumber and cowpea matrices.

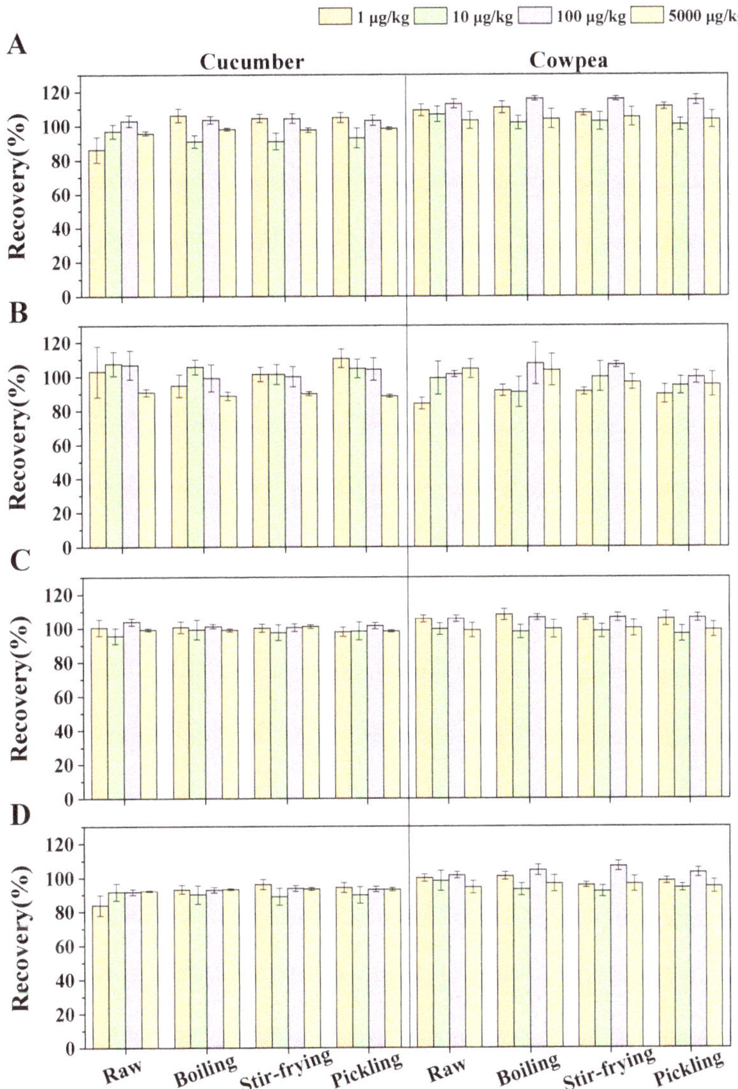

Figure 2. Recoveries of FLU (**A**) and TRI (**B**) and their metabolites, FLB (**C**) and TRA (**D**), in different matrices. Note: The error bars represent standard deviations calculated from five duplicates.

3.2. Residue Dissipation and Accumulation of FLU and TRI in Greenhouse Cucumbers and Cowpeas

In our study, cucumbers and cowpeas were sprayed with FLU and TRI three consecutive times every seven days; all the residue data and dissipation curves are listed in Table 1 and Figure 3. The FLU residues in cucumbers ranged from 21.44 µg/kg to 107.13 µg/kg following the first spraying event, 120.33 µg/kg to 225.89 µg/kg following the second spraying event and 152.00 µg/kg to 247.65 µg/kg following the third spraying event, and those in cowpeas were 877.37–1341.08 µg/kg, 1626.35–2341.41 µg/kg and 2103.66–3256.29 µg/kg, respectively. These results suggest that cowpeas had much higher levels of FLU residues than cucumbers. A similar residue pattern was found between TRI levels in cucumbers and cowpeas (Table 1). FLU and TRI residues were then used to fit the first-order kinetics models, which produced satisfactory correlation coefficients of 0.50–0.97. The calculated half-lives of FLU were 3.15, 8.89 and 10.66 days for cucumbers and 12.37, 16.91 and 15.40 days for cowpeas, and those for TRI were 2.60, 4.88 and 7.97 days for cucumbers and 17.33, 22.36 and 10.83 days for cowpeas. These results suggest that FLU dissipated faster in cucumbers than in cowpeas, whereas TRI showed a similar dissipation trend in cucumbers and cowpeas. In addition, we noted that residual FLU and TRI in cucumbers and cowpeas increased with increasing spraying times. The average RA values of FLU were 1:2.65:1.08 for cucumbers and 1:1.83:1.15 for cowpeas, while those for TRI were 1:2.04:1.25 for cucumbers and 1:1.69:1.17 for cowpeas.

Table 1. Residues, dissipation kinetics and residue accumulation (RA) of FLU and TRI and their metabolites, FLB and TRA, in cucumbers and cowpeas after repeated spraying. Note: SD denotes standard deviation, calculated from three duplicates.

Time	1st Spraying		2nd Spraying		3rd Spraying		RA	Average RA
	Mean Residue $\pm$ SD (µg/kg)	Dissipation Rate (%)	Mean Residue $\pm$ SD (µg/kg)	Dissipation Rate (%)	Mean Residue $\pm$ SD (µg/kg)	Dissipation Rate (%)		
FLU in cucumbers								
2 h	105.63 $\pm$ 2.14	-	225.89 $\pm$ 3.59	-	214.67 $\pm$ 2.73	-	1:2.14:0.95	
1 d	107.13 $\pm$ 2.68	-	195.11 $\pm$ 12.11	13.63	247.65 $\pm$ 6.16	-	1:1.82:1.27	
3 d	54.78 $\pm$ 2.03	48.14	197.11 $\pm$ 26.85	12.74	209.42 $\pm$ 4.04	2.44	1:3.60:1.06	1:2.65:1.08
5 d	54.27 $\pm$ 1.10	48.62	172.08 $\pm$ 9.93	23.82	152.00 $\pm$ 2.25	29.19	1:3.17:0.88	
7 d	21.44 $\pm$ 0.62	79.70	120.33 $\pm$ 1.03	46.73	157.22 $\pm$ 2.77	26.76	1:5.61:1.31	
Dynamic equation	$y = 120.05e^{-0.22x}$		$y = 228.98e^{-0.078x}$		$y = 237.76e^{-0.065x}$			
R^2	0.90		0.85		0.77			
t1/2 (d)	3.15		8.89		10.66			
FLU in cowpeas								
2 h	1341.08 $\pm$ 53.95	-	2341.41 $\pm$ 18.12	-	3256.29 $\pm$ 19.45	-	1:1.75:1.39	
1 d	1215.85 $\pm$ 27.27	9.34	2142.03 $\pm$ 38.70	8.52	2228.24 $\pm$ 79.55	31.57	1:1.76:1.04	
3 d	1179.58 $\pm$ 39.45	12.04	2131.32 $\pm$ 29.49	8.97	2207.41 $\pm$ 37.26	32.21	1:1.81:1.04	1:1.83:1.15
5 d	1049.90 $\pm$ 5.85	21.71	2130.55 $\pm$ 41.46	9.01	2160.21 $\pm$ 30.58	33.66	1:2.03:1.01	
7 d	877.37 $\pm$ 28.36	34.58	1626.35 $\pm$ 7.56	30.54	2103.66 $\pm$ 49.25	35.40	1:1.85:1.29	
Dynamic equation	$y = 1341.06e^{-0.056x}$		$y = 2351.11e^{-0.041x}$		$y = 2726.41e^{-0.045x}$			
R^2	0.95		0.72		0.50			
t1/2	12.37		16.91		15.40			
TRI in cucumbers								
2 h	110.14 $\pm$ 8.39	-	207.41 $\pm$ 13.18	-	202.85 $\pm$ 11.54	-	1:1.88:0.98	
1 d	109.14 $\pm$ 9.53	0.91	148.30 $\pm$ 1.86	28.50	208.68 $\pm$ 17.17	-	1:1.36:1.41	
3 d	49.54 $\pm$ 2.92	55.02	162.21 $\pm$ 4.13	21.79	173.55 $\pm$ 24.85	14.45	1:3.27:1.07	1:2.04:1.25
5 d	44.04 $\pm$ 1.13	60.02	67.62 $\pm$ 0.19	67.40	135.30 $\pm$ 8.18	33.30	1:1.54:2.00	
7 d	16.48 $\pm$ 0.97	85.03	85.18 $\pm$ 4.68	58.93	117.59 $\pm$ 10.56	42.03	1:5.17:1.38	
Dynamic equation	$y = 126.06e^{-0.28x}$		$y = 194.74e^{-0.14x}$		$y = 216.45e^{-0.087x}$			
R^2	0.94		0.74		0.97			
t1/2 (d)	2.60		4.88		7.97			
TRI in cowpeas								
2 h	1586.16 $\pm$ 156.66	-	2627.72 $\pm$ 124.51	-	3576.15 $\pm$ 79.16	-	1:1.66:1.36	
1 d	1384.96 $\pm$ 24.75	12.68	2200.96 $\pm$ 52.53	16.24	2665.77 $\pm$ 80.46	25.46	1:1.59:1.21	
3 d	1235.94 $\pm$ 29.10	22.08	2187.77 $\pm$ 99.08	16.74	2399.93 $\pm$ 48.44	32.89	1:1.77:1.1	1:1.69:1.17
5 d	1267.11 $\pm$ 60.87	20.11	2194.10 $\pm$ 49.65	16.50	2258.87 $\pm$ 33.36	36.84	1:1.73:1.03	
7 d	1138.03 $\pm$ 131.06	28.25	1964.60 $\pm$ 75.54	25.24	2122.85 $\pm$ 79.55	40.64	1:1.73:1.08	
Dynamic equation	$y = 1495.90\,e^{-0.040x}$		$y = 2455.67e^{-0.031x}$		$y = 3140.45e^{-0.064x}$			
R^2	0.82		0.69		0.78			
t1/2	17.33		22.36		10.83			

Table 1. *Cont.*

Time	1st Spraying		2nd Spraying		3rd Spraying		RA	Average RA
	Mean Residue $\pm$ SD (μg/kg)	Dissipation Rate (%)	Mean Residue $\pm$ SD (μg/kg)	Dissipation Rate (%)	Mean Residue $\pm$ SD (μg/kg)	Dissipation Rate (%)		
FLB in cowpeas								
2 h	<LOQ		1.58 $\pm$ 0.04		3.91 $\pm$ 0.19		-	
1 d	<LOQ		2.03 $\pm$ 0.07		3.93 $\pm$ 0.30		-	
3 d	<LOQ		2.61 $\pm$ 0.14		6.40 $\pm$ 0.09		-	1:4.01:1.99
5 d	<LOQ		2.27 $\pm$ 0.08		8.33 $\pm$ 0.18		-	
7 d	1.14 $\pm$ 0.14		4.57 $\pm$ 0.35		9.06 + 0.31		1:4.01:1.99	
TRA in cucumbers								
2 h	1.69 $\pm$ 0.11		5.57 $\pm$ 0.09		20.30 $\pm$ 0.57		1:3.29:3.64	
1 d	4.27 $\pm$ 0.01		9.53 $\pm$ 0.32		24.26 $\pm$ 0.57		1:2.23:2.55	
3 d	7.81 $\pm$ 0.27		15.27 $\pm$ 2.69		19.54 $\pm$ 0.54		1:1.96:1.28	1:2.71:1.53
5 d	3.96 $\pm$ 0.07		11.64 $\pm$ 1.15		8.40 $\pm$ 0.17		1:2.94:0.72	
7 d	1.89 $\pm$ 0.15		11.16 $\pm$ 0.06		8.76 $\pm$ 0.11		1:5.92:0.79	
TRA in cowpeas								
2 h	5.85 $\pm$ 0.33		49.52 $\pm$ 1.48		35.79 $\pm$ 1.22		1:8.47:0.72	
1 d	16.42 $\pm$ 1.78		54.21 $\pm$ 4.50		57.19 $\pm$ 6.12		1:3.30:1.05	
3 d	15.34 $\pm$ 0.28		55.26 $\pm$ 4.77		76.17 $\pm$ 0.68		1:4.97:0.73	1:2.86:1.14
5 d	20.85 $\pm$ 0.30		55.05 $\pm$ 0.84		65.16 $\pm$ 3.18		1:2.64:1.18	
7 d	29.71 $\pm$ 8.50		37.98 $\pm$ 0.75		51.89 $\pm$ 4.48		1:1.28:1.37	

The parent FLU and TRI were the main compounds present in the greenhouse-field cucumbers and cowpeas, and their metabolites (FLB and TRA) fluctuated at low residue levels. Residual FLB was only found in cowpeas, with detectable concentrations of 1.14–9.06 μg/kg. Residual TRA ranged from 1.69 μg/kg to 24.26 μg/kg in cucumbers and from 5.85 μg/kg to 76.17 μg/kg in cowpeas, indicating higher levels of TRA residues in cowpeas (Table 1). In general, the concentrations of TRA first increased and then gradually decreased 2 h–7 d after each spraying event, except for the first instance of spraying on cowpeas, whereby TRA concentrations underwent a sustained increase (Figure 3). Although the levels of FLB and TRA residues were low, both showed strong residue accumulation in cucumbers (average RA value of TRA, 1:2.71:1.53) and cowpeas (FLB, 1:4.01:1.99; TRA, 1:2.86:1.14). Together, these results verify the residue accumulation of parent FLU and TRI and their metabolites, FLB and TRA, in greenhouse cucumbers and cowpeas after repeated spraying.

3.3. Residue Fate and Processing Factors of FLU and TRI in Cucumbers and Cowpeas

Pesticide residue changes in cucumbers and cowpeas were studied during several traditional household processing operations, including peeling, washing, stir-frying, boiling and pickling; all the data are listed in Table S4. Pesticide residue reduction and concentration during processing were estimated using the PFs, which are shown in Figure 4. Peeling removed 70.06% of FLU, 76.40% of TRI and 86.80% of TRA from cucumbers (Table S4). Following washing, the PFs in cucumbers decreased from 0.77 (1 min) to 0.38 (10 min) for FLU and 0.90 (1 min) to 0.30 (10 min) for TRI. A similar stepped downward trend was found for PFs in cowpeas for both FLU and TRI following washing (Figure 4A). Following boiling, the PFs of FLU and TRI also presented a decreasing trend in cucumbers and cowpeas with increasing in processing time (Figure 4C). However, following stir-frying, the PFs of FLU increased from 0.29 to 0.58 for cucumbers and 0.61 to 0.74 for cowpeas in 10 min, and those for TRI increased from 0.12 to 0.47 for cucumbers and 0.55 to 0.77 for cowpeas (Figure 4B). Following pickling, the PFs of FLU and TRI first decreased and then gradually increased (Figure 4D), implying that overpickling may result in an increase in pesticide residues in pickled cucumbers and cowpeas. Although the pesticide residues fluctuated greatly during processing, all operations reduced FLU and TRI residues in cucumbers and cowpeas because their PFs were lower than 1.

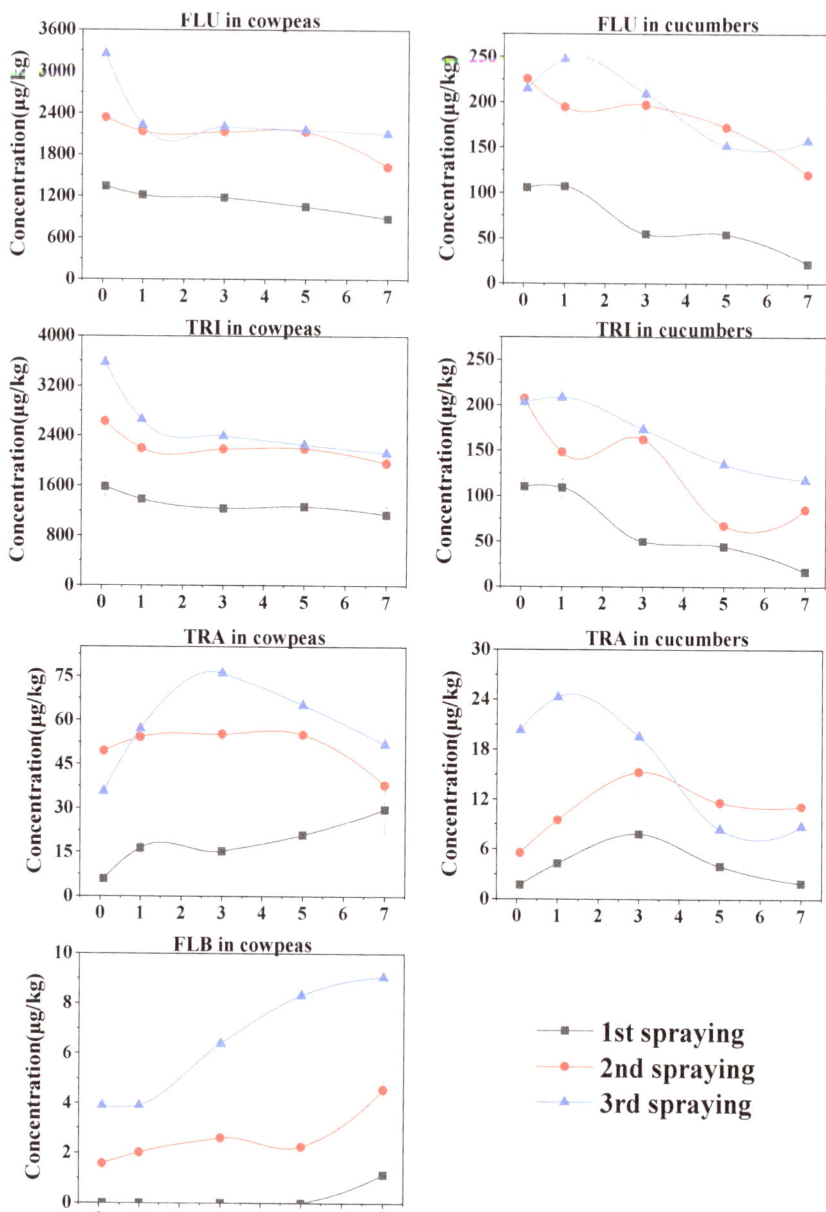

Figure 3. Residue dissipation of FLU and TRI and their metabolites, FLB and TRA, in cucumbers and cowpeas after repeated spraying. Note: The error bars represent standard deviations calculated from three duplicates.

In addition to the parent FLU and TRI, the residues of FLB and TRA changed greatly in the processed cucumbers and cowpeas (Table S4). Washing, stir-frying and boiling removed large amounts of TRA from cucumbers and cowpeas with PFs of 0.19–0.90, except for cucumbers washed for 1 min (1.11); on the contrary, pickling increased the concentration of TRA residues in cucumbers (PF range, 1.39–2.73) and cowpeas (PF range, 1.35–5.41). Residual FLB was low in the processed cowpeas at 1.41–6.91 µg/kg, and PF values of higher than 1 were found for stir-frying for 7 min (1.17) and 10 min (1.23) and boiling for 5 min (1.17) and 7 min (1.23). In total, more attention should be paid to pickling operations due to their obvious effects on TRA concentrations during processing.

Figure 4. Processing factors (PFs) of FLU and TRI and their metabolites, FLB and TRA, in cucumbers and cowpeas following washing (**A**), stir-frying (**B**), boiling (**C**) and pickling (**D**). The red bars represent a PF value of >1.

3.4. Dietary Risk Assessment of FLU and TRI in Cucumbers and Cowpeas

The maximum residue limits (MRLs) were first used to estimate the residue risks of FLU and TRI in cucumbers and cowpeas. The total FLU and TRI residues (expressed as FLU_{sum} and TRI_{sum}, respectively) at different sampling intervals are listed in Table S5. A 3-day pre-harvest interval (PHI) for the 43% suspension concentrate of FLU and TRI was proposed for both cucumbers and cowpeas. In cucumbers, even the highest residues of FLU_{sum} (253.21 µg/kg) and TRI_{sum} (277.61 µg/kg) were less than the MRLs recommended by China, the CAC, the US, the EU, Japan and Korea (500–1000 µg/kg for FLU and 300–700 µg/kg for TRI). In cowpeas, the average FLU_{sum} residues 3 days after each spraying event ranged from 1180.10 to 2214.06 µg/kg and were higher than the MRLs of China and CAC (1000 µg/kg) but lower than the MRLs of the US and the EU (3000–4000 µg/kg). Moreover, the average TRI_{sum} residue 3 days after the first spraying event (1251.89 µg/kg) was lower than the MRLs of the US and the EU (1500 µg/kg), whereas those following the second and third spraying events (2245.24 µg/kg and 2479.15 µg/kg) were higher than these MRLs.

The chronic and acute risks associated with exposure to FLU and TRI from cucumbers and cowpeas were estimated based on FAO/WHO models; the risk values are listed in Table 2. For the chronic risk assessment, the HQc values of FLU for children and adults ranging from 0.0033 to 0.042 for cucumbers and 0.016 to 0.099 for cowpeas, and those for TRI were 0.0010–0.011 for cucumbers and 0.016–0.099 for cowpeas, which were all far below 1. For the acute risk assessment, the HQa values ranged from 0.0027 to 0.017 for cucumbers and 0.023 to 0.057 for cowpeas for both children and adults, i.e., <1. In addition, acute risk assessment for TRI was not considered as the ARfD value was reported to be unnecessary by JMPR and EFSA [32,33]. In conclusion, the human health risks of dietary exposure to FLU and TRI from cucumber and cowpea consumption were considered to be acceptable based on the results of the present study.

Table 2. Chronic and acute hazard quotients of FLU and TRI in cucumbers and cowpeas for children and adults.

Pesticides	Vegetables	Spraying Time	NEDI (μg/kg bw/d)		HQc		IESTI (μg/kg bw/d)		HQa	
			Children	Adults	Children	Adults	Children	Adults	Children	Adults
FLU	Cucumber	1	0.11	0.033	0.011	0.0033	2.26	1.37	0.0045	0.0027
		2	0.39	0.12	0.039	0.012	8.72	5.29	0.017	0.011
		3	0.42	0.13	0.042	0.013	8.46	5.14	0.017	0.010
	Cowpea	1	0.52	0.16	0.052	0.016	15.50	11.73	0.031	0.023
		2	0.95	0.29	0.095	0.029	27.41	20.75	0.055	0.041
		3	0.99	0.30	0.099	0.030	28.33	21.45	0.057	0.043
TRI	Cucumber	1	0.13	0.040	0.0033	0.0010	2.73	1.66	-	-
		2	0.38	0.12	0.010	0.0029	8.06	4.90	-	-
		3	0.43	0.13	0.011	0.0032	9.39	5.70	-	-
	Cowpea	1	0.57	0.17	0.014	0.0043	16.02	12.13	-	-
		2	1.00	0.30	0.025	0.0076	29.81	22.56	-	-
		3	1.11	0.34	0.028	0.0084	31.77	24.05	-	-

4. Discussion

Studying the dissipation, metabolism and accumulation of pesticides after repeated spraying is of vital importance to ensure food safety. In our study, FLU and TRI residues were higher in cowpeas than in cucumbers. The residue difference between cucumbers and cowpeas may be largely due to differences in the crops' morphological characteristics, including crop height, foliar size and shape, fruit size, shape and surface [34]. Moreover, FLU and TRI residues dissipated faster in cucumbers (half-life range, 2.60–10.66 days) than in cowpeas (half-life range, 10.83–22.36 days). During the experimental period, cucumbers grew rapidly to contribute to the dilution of pesticide residues in fruits; however, cowpeas only showed a slight increase in fruit weight after pesticide spraying. This may be the main reason for the fast dissipation of FLU and TRI in cucumbers. Moreover, the half-lives of cowpeas and cucumbers in our study were larger than those in the chili (1.23–2.38 for FLU; 1.27–4.88 for TRI) and onion (1.74–1.83 for FLU; 4.73–4.78 for TRI) [9,12]. Our results suggest that both FLU and TRI showed potential residue accumulation in cucumbers (average RA value, 1:2.65:1.08 for FLU and 1:2.04:1.25 for TRI) and cowpeas (average RA value, 1:1.83:1.15 for FLU and 1:1.69:1.17 for TRI) after repeated spraying. Similarly, Wang et al. confirmed the residue accumulation of four common fungicides in strawberries in the order of procymidone > cyprodinil > pyrimethanil > pyraclostrobin after repeated spraying [14]. In addition to the parent compounds, their metabolites, FLB and TRA, also accumulated in cucumbers (average RA values of TRA, 1:2.71:1.53) and cowpeas (FLB, 1:4.01:1.99; TRA, 1:2.86:1.14). Other studies have also confirmed the prevalence of FLB and TRA in crops [12,13,35]. Therefore, due attention should be given to these metabolites to achieve a comprehensive risk assessment of FLU and TRI in cucumbers and cowpeas.

Food processing is typically accompanied by pesticide residue changes. In our study, peeling, washing, stir-frying, boiling and pickling partially or substantially removed FLU and TRI residues from raw agricultural products (PF range, 0.12–0.97). Unlike washing and boiling operations, residues of FLU and TRI presented an increasing trend with increasing stir-frying and pickling time. We inferred that the main reason for this residue increase was the water loss that occurs in cucumbers and cowpeas under high-temperature conditions (stir-frying) and in high-salt environments (pickling). Similarly, Lu et al. found that residues of organophosphorus pesticides, such as chlorpyrifos, isocarbophos, profenofos and triazophos, were also increased to some extent during the cabbage pickling process with different NaCl solutions [36]. Huan et al. found that residual procymidone, chlorothalonil and difenoconazole increased 1.03–1.11 times in stir-fried cowpeas [27]. More notably, residues of TRA appeared to be concentrated in pickled cucumbers and cowpeas (PF range, 1.35–5.41), and these concentration effects increased with increasing pickling time. TRI was easily hydrolyzed into TRA via cleavage of its methyl ester group in aquatic environments [37,38].

The health risks of chronic and acute exposure to total FLU and TRI from cucumbers and cowpeas were assessed for both children and adults. The HQc and HQa values of total FLU and TRI ranged from 0.0010 to 0.099 and 0.0027 to 0.057, respectively, although the highest value was still far smaller than 1, indicating low risks of adverse effects following exposure to FLU and TRI based on the results of the present study. Similarly, previous studies also revealed a low dietary risk of exposure to FLU or TRI from various agricultural products, such as onions, carrots, strawberries and mangoes [12,39–41]. The risks associated with exposure to FLU and TRI should not be ignored. On the one hand, FLU and TRI residue concentrations were high in cucumbers and cowpeas, and some residues at the PHI even exceeded the MRLs established by the Chinese government. On the other hand, long-term exposure to low doses of pesticides (even at permitted levels) could have harmful effects on mammals [42–44]. Moreover, numerous studies have confirmed the cumulative toxicity of different pesticides that were classified in both the same and different classes [31,45,46]. Therefore, individual or mixed processing operations should be used to reduce FLU and TRI residues in agricultural products, with the aim of protecting human health.

5. Conclusions

In this study, we comprehensively assessed the dissipation, metabolism, accumulation, processing and risk assessment of FLU and TRI in cucumbers and cowpeas from cultivation to consumption for the first time. Our results suggest that FLU and TRI residues were higher in cowpeas but dissipated faster in cucumbers. FLU and TRI were the main compounds found in field samples ($\leq$3256.29 µg/kg), and their metabolites (FLB and TRA) fluctuated at low residue levels ($\leq$76.17 µg/kg). Moreover, both FLU and TRI accumulated in cucumbers and cowpeas after repeated spraying. Our results show that peeling, washing, stir-frying, boiling and pickling could partially or substantially remove FLU and TRI residues from raw cucumbers and cowpeas; on the contrary, residues of the metabolite TRA were obviously concentrated following pickling. Chronic and acute risk assessments indicate that exposure to FLU and TRI through cucumber and cowpea consumption posed a low health risk to either children or adults based on the results of the present study. In the future, more experimental sites and crop categories across China should be chosen to study the fate of FLU and TRI residues to achieve a comprehensive understanding of their actual dietary risks; this would be a significant step in ensuring the safe use of FLU- and TRI-containing products and in protecting human health.

Supplementary Materials: The following supporting information can be downloaded at: https://www.mdpi.com/article/10.3390/foods12102082/s1, Table S1: Retention time and MRM parameters of FLU and TRI and their metabolites FLB and TRA for HPLC-MS/MS detection; Table S2: Detailed parameters/exposure factors for human health risk assessment; Table S3: The calibration regression equation, R2, ME, and LOQ for FLU and TRI and their metabolites FLB and TRA in different matrices; Table S4: Residue levels of FLU and TRI and their metabolites FLB and TRA in cucumbers and cowpeas for the different processing operations; Table S5: Total FLU and TRI residues (expressed as FLUsum and TRIsum) at different sampling intervals and MRLs for different countries.

Author Contributions: Conceptualization, K.C. and Z.D.; methodology, S.G.; software, J.L. and L.F.; validation, J.L. and L.F.; formal analysis, K.C. and R.D.; investigation, J.W.; resources, T.L.; data curation, Z.D.; writing—original draft preparation, K.C.; writing—review and editing, X.W. and Y.Z.; visualization, K.C.; supervision, Z.D.; project administration, Z.D.; funding acquisition, K.C. and Z.D. All authors have read and agreed to the published version of the manuscript.

Funding: This research was funded by the Agricultural Scientific and Technological Innovation Project of Shandong Academy of Agricultural Sciences (No. CXGC2022E05 and CXGC2023044) and the Open Project Program of Shandong Provincial Key Laboratory of Test Technology on Food Quality and Safety (No. 2021KF01).

Data Availability Statement: The data presented in this study are available on request from the corresponding author.

Conflicts of Interest: All authors declare no competing financial interest.

References

1. Cui, K.; Ning, M.; Liang, J.; Guan, S.; Fang, L.; Ding, R.; Wang, J.; Li, T.; Dong, Z. Pollution characteristics and non-dietary human cumulative risk assessment of neonicotinoids in vegetable greenhouse soils: A case study in Shandong Province, China. *J. Soils Sediments* **2023**, *23*, 331–343. [CrossRef]
2. Ezrari, S.; Lazraq, A.; El Housni, Z.; Radouane, N.; Belabess, Z.; Mokrini, F.; Tahiri, A.; Amiri, S.; Lahlali, R. Evaluating the sensitivity and efficacy of fungicides with different modes of action against Neocosmospora solani and Fusarium species, causing agents of citrus dry root rot. *Arch. Phytopathol. Plant Prot.* **2022**, *55*, 1117–1135. [CrossRef]
3. Rouquié, D.; Tinwell, H.; Blanck, O.; Schorsch, F.; Geter, D.; Wason, S.; Bars, R. Thyroid tumor formation in the male mouse induced by fluopyram is mediated by activation of hepatic CAR/PXR nuclear receptors. *Regul. Toxicol. Pharmacol.* **2014**, *70*, 673–680. [CrossRef] [PubMed]
4. Tinwell, H.; Rouquié, D.; Schorsch, F.; Geter, D.; Wason, S.; Bars, R. Liver tumor formation in female rat induced by fluopyram is mediated by CAR/PXR nuclear receptor activation. *Regul. Toxicol. Pharmacol.* **2014**, *70*, 648–658. [CrossRef] [PubMed]
5. Jang, Y.; Lee, A.Y.; Chang, S.H.; Jeong, S.H.; Park, K.H.; Paik, M.K.; Cho, N.J.; Kim, J.E.; Cho, M.H. Trifloxystrobin induces tumor necrosis factor-related apoptosis-inducing ligand (TRAIL)-mediated apoptosis in HaCaT, human keratinocyte cells. *Drug Chem. Toxicol.* **2017**, *40*, 67–73. [CrossRef]
6. Nguyen, K.; Sanchez, C.L.; Brammer-Robbins, E.; Pena-Delgado, C.; Kroyter, N.; El Ahmadie, N.; Watkins, J.M.; Aristizabal-Henao, J.J.; Bowden, J.A.; Souders, C.L., II. Neurotoxicity assessment of QoI strobilurin fungicides azoxystrobin and trifloxystrobin in human SH-SY5Y neuroblastoma cells: Insights from lipidomics and mitochondrial bioenergetics. *Neurotoxicology* **2022**, *91*, 290–304. [CrossRef]
7. Cao, M.; Li, S.; Wang, Q.; Wei, P.; Liu, Y.; Zhu, G.; Wang, M. Track of fate and primary metabolism of trifloxystrobin in rice paddy ecosystem. *Sci. Total Environ.* **2015**, *518*, 417–423. [CrossRef]
8. Chawla, S.; Patel, D.J.; Patel, S.H.; Kalasariya, R.L.; Shah, P.G. Behaviour and risk assessment of fluopyram and its metabolite in cucumber (*Cucumis sativus*) fruit and in soil. *Environ. Sci. Pollut. Res.* **2018**, *25*, 11626–11634. [CrossRef]
9. Mandal, K.; Singh, R.; Sharma, S.; Kataria, D. Dissipation and kinetic studies of fluopyram and trifloxystrobin in chilli. *J. Food Compost. Anal.* **2023**, *115*, 105008. [CrossRef]
10. Matadha, N.Y.; Mohapatra, S.; Siddamallaiah, L. Distribution of fluopyram and tebuconazole in pomegranate tissues and their risk assessment. *Food Chem.* **2021**, *358*, 129909. [CrossRef]
11. Mohapatra, S. Residue levels and dissipation behaviors for trifloxystrobin and tebuconazole in mango fruit and soil. *Environ. Monit. Assess.* **2015**, *187*, 95. [CrossRef] [PubMed]
12. Sharma, N.; Mandal, K.; Sharma, S. Dissipation and risk assessment of fluopyram and trifloxystrobin on onion by GC–MS/MS. *Environ. Sci. Pollut. Res.* **2022**, *29*, 80612–80623. [CrossRef]
13. Wang, L.; Li, W.; Li, P.; Li, M.; Chen, S.; Han, L. Residues and dissipation of trifloxystrobin and its metabolite in tomatoes and soil. *Environ. Monit. Assess.* **2014**, *186*, 7793–7799. [CrossRef] [PubMed]
14. Wang, Z.; Di, S.; Qi, P.; Xu, H.; Zhao, H.; Wang, X. Dissipation, accumulation and risk assessment of fungicides after repeated spraying on greenhouse strawberry. *Sci. Total Environ.* **2021**, *758*, 144067. [CrossRef] [PubMed]
15. Arisekar, U.; Shakila, R.J.; Shalini, R.; Jeyasekaran, G.; Padmavathy, P. Effect of household culinary processes on organochlorine pesticide residues (OCPs) in the seafood (*Penaeus vannamei*) and its associated human health risk assessment: Our vision and future scope. *Chemosphere* **2022**, *297*, 134075. [CrossRef] [PubMed]
16. Li, Y.; Xu, J.; Zhao, X.; He, H.; Zhang, C.; Zhang, Z. The dissipation behavior, household processing factor and risk assessment for cyenopyrafen residues in strawberry and mandarin fruits. *Food Chem.* **2021**, *359*, 129925. [CrossRef] [PubMed]
17. Kim, S.W.; Abd El-Aty, A.; Rahman, M.M.; Choi, J.H.; Lee, Y.J.; Ko, A.Y.; Choi, O.J.; Jung, H.N.; Hacımüftüoğlu, A.; Shim, J.H. The effect of household processing on the decline pattern of dimethomorph in pepper fruits and leaves. *Food Control* **2015**, *50*, 118–124. [CrossRef]

18. Yang, A.; Park, J.H.; Abd El-Aty, A.; Choi, J.H.; Oh, J.H.; Do, J.A.; Kwon, K.; Shim, K.H.; Choi, O.J.; Shim, J.H. Synergistic effect of washing and cooking on the removal of multi-classes of pesticides from various food samples. *Food Control* **2012**, *28*, 99–105. [CrossRef]
19. Yigit, N.; Velioglu, Y.S. Effects of processing and storage on pesticide residues in foods. *Crit. Rev. Food Sci. Nutr.* **2020**, *60*, 3622–3641. [CrossRef]
20. Słowik-Borowiec, M.; Szpyrka, E. Selected food processing techniques as a factor for pesticide residue removal in apple fruit. *Environ. Sci. Pollut. Res.* **2020**, *27*, 2361–2373. [CrossRef]
21. Noh, H.H.; Kwon, H.; Lee, H.S.; Ro, J.; Kyung, K.S.; Moon, B. Determination of pyraclostrobin and trifloxystrobin residues in red pepper powder processed from raw red pepper. *Food Anal. Methods* **2019**, *12*, 94–99. [CrossRef]
22. Kong, Z.; Quan, R.; Fan, B.; Liao, Y.; Chen, J.; Li, M.; Dai, X. Stereoselective behaviors of the fungicide triadimefon and its metabolite triadimenol during malt storage and beer brewing. *J. Hazard. Mater.* **2020**, *400*, 123238. [CrossRef] [PubMed]
23. Li, H.; Zhong, Q.; Wang, X.; Luo, F.; Zhou, L.; Sun, H.; Yang, M.; Lou, Z.; Chen, Z.; Zhang, X. The degradation and metabolism of chlorfluazuron and flonicamid in tea: A risk assessment from tea garden to cup. *Sci. Total Environ.* **2021**, *754*, 142070. [CrossRef] [PubMed]
24. Li, C.; Zhu, H.; Li, C.; Qian, H.; Yao, W.; Guo, Y. The present situation of pesticide residues in China and their removal and transformation during food processing. *Food Chem.* **2021**, *354*, 129552. [CrossRef] [PubMed]
25. Zhao, L.; Liu, F.; Wu, L.; Xue, X.; Hou, F. Fate of triadimefon and its metabolite triadimenol in jujube samples during jujube wine and vinegar processing. *Food Control* **2017**, *73*, 468–473. [CrossRef]
26. An, X.; Pan, X.; Li, R.; Jiang, D.; Dong, F.; Zhu, W.; Xu, J.; Liu, X.; Wu, X.; Zheng, Y. Enantioselective monitoring chiral fungicide mefentrifluconazole in tomato, cucumber, pepper and its pickled products by supercritical fluid chromatography tandem mass spectrometry. *Food Chem.* **2022**, *376*, 131883. [CrossRef] [PubMed]
27. Huan, Z.; Xu, Z.; Jiang, W.; Chen, Z.; Luo, J. Effect of Chinese traditional cooking on eight pesticides residue during cowpea processing. *Food Chem.* **2015**, *170*, 118–122. [CrossRef]
28. FAO/WHO. Global Environment Monitoring System (GEMS)/Food Contamination Monitoring and Assessment Programme. 2023. Available online: https://www.who.int/teams/nutrition-and-food-safety/databases/global-environment-monitoring-system-food-contamination (accessed on 20 January 2023).
29. Zhou, J.; Dong, C.; An, W.; Zhao, Q.; Zhang, Y.; Li, Z.; Jiao, B. Dissipation of imidacloprid and its metabolites in Chinese prickly ash (Zanthoxylum) and their dietary risk assessment. *Ecotoxicol. Environ. Saf.* **2021**, *225*, 112719. [CrossRef]
30. Wang, X.; Zhang, X.; Wang, Z.; Zhou, L.; Luo, F.; Chen, Z. Dissipation behavior and risk assessment of tolfenpyrad from tea bushes to consuming. *Sci. Total Environ.* **2022**, *806*, 150771. [CrossRef]
31. Cui, K.; Wu, X.; Wei, D.; Zhang, Y.; Cao, J.; Xu, J.; Dong, F.; Liu, X.; Zheng, Y. Health risks to dietary neonicotinoids are low for Chinese residents based on an analysis of 13 daily-consumed foods. *Environ. Int.* **2021**, *149*, 106385. [CrossRef]
32. European Food Safety Authority (EFSA); Arena, M.; Auteri, D.; Barmaz, S.; Bellisai, G.; Brancato, A.; Brocca, D.; Bura, L.; Byers, H.; Chiusolo, A.; et al. Peer review of the pesticide risk assessment of the active substance trifloxystrobin. *EFSA J.* **2017**, *15*, e04989.
33. JMPR. Joint FAO/WHO Meeting on Pesticide Residues. *Trifloxystrobin. JMPR Report.* 2017. Available online: https://www.fao.org/fileadmin/templates/agphome/documents/Pests_Pesticides/JMPR/Report2017/5.37_TRIFLOXYSTROBIN__213_.pdf (accessed on 20 January 2023).
34. Noh, H.H.; Lee, J.Y.; Park, H.K.; Lee, J.W.; Jo, S.H.; Lim, J.B.; Shin, H.G.; Kwon, H.; Kyung, K.S. Dissipation, persistence, and risk assessment of fluxapyroxad and penthiopyrad residues in perilla leaf (*Perilla frutescens var. japonica Hara*). *PLoS ONE* **2019**, *14*, e0212209. [CrossRef]
35. Vargas-Pérez, M.; Egea González, F.J.; Garrido Frenich, A. Dissipation and residue determination of fluopyram and its metabolites in greenhouse crops. *J. Sci. Food Agric.* **2020**, *100*, 4826–4833. [CrossRef]
36. Lu, Y.; Yang, Z.; Shen, L.; Liu, Z.; Zhou, Z.; Diao, J. Dissipation behavior of organophosphorus pesticides during the cabbage pickling process: Residue changes with salt and vinegar content of pickling solution. *J. Agric. Food Chem.* **2013**, *61*, 2244–2252. [CrossRef]
37. Chen, X.; Xu, J.; Liu, X.; Tao, Y.; Pan, X.; Zheng, Y.; Dong, F. Simultaneous determination of trifloxystrobin and trifloxystrobin acid residue in rice and soil by a modified quick, easy, cheap, effective, rugged, and safe method using ultra high performance liquid chromatography with tandem mass spectrometry. *J. Sep. Sci.* **2014**, *37*, 1640–1647. [CrossRef]
38. Bian, Y.; Feng, Y.; Zhang, A.; Qi, X.; Ma, X.; Pan, J.; Han, J.; Liang, L. Residue behaviors, processing factors and transfer rates of pesticides and metabolites in rose from cultivation to consumption. *J. Hazard. Mater.* **2023**, *442*, 130104. [CrossRef] [PubMed]
39. Mohapatra, S.; Siddamallaiah, L.; Buddidathi, R.; Yogendraiah Matadha, N. Dissipation kinetics and risk assessment of fluopyram and tebuconazole in mango (*Mangifera indica*). *Int. J. Environ. Anal. Chem.* **2018**, *98*, 229–246. [CrossRef]
40. Song, L.; Zhong, Z.; Han, Y.; Zheng, Q.; Qin, Y.; Wu, Q.; He, X.; Pan, C. Dissipation of sixteen pesticide residues from various applications of commercial formulations on strawberry and their risk assessment under greenhouse conditions. *Ecotoxicol. Environ. Saf.* **2020**, *188*, 109842. [CrossRef]
41. Yang, Y.; Yang, M.; Zhao, T.; Pan, L.; Jia, L.; Zheng, L. Residue and Risk Assessment of Fluopyram in Carrot Tissues. *Molecules* **2022**, *27*, 5544. [CrossRef] [PubMed]

42. Conley, J.M.; Lambright, C.S.; Evans, N.; Cardon, M.; Medlock-Kakaley, E.; Wilson, V.S.; Gray, L.E., Jr. A mixture of 15 phthalates and pesticides below individual chemical no observed adverse effect levels (NOAELs) produces reproductive tract malformations in the male rat. *Environ. Int.* **2021**, *156*, 106615. [CrossRef] [PubMed]

43. Ghasemnejad-Berenji, M.; Nemati, M.; Pourheydar, B.; Gholizadeh, S.; Karimipour, M.; Mohebbi, I.; Jafari, A. Neurological effects of long-term exposure to low doses of pesticides mixtures in male rats: Biochemical, histological, and neurobehavioral evaluations. *Chemosphere* **2021**, *264*, 128464. [CrossRef] [PubMed]

44. Tsatsakis, A.; Docea, A.O.; Constantin, C.; Calina, D.; Zlatian, O.; Nikolouzakis, T.K.; Stivaktakis, P.D.; Kalogeraki, A.; Liesivuori, J.; Tzanakakis, G. Genotoxic, cytotoxic, and cytopathological effects in rats exposed for 18 months to a mixture of 13 chemicals in doses below NOAEL levels. *Toxicol. Lett.* **2019**, *316*, 154–170. [CrossRef] [PubMed]

45. Alarcan, J.; Waizenegger, J.; Solano, M.; Lichtenstein, D.; Luckert, C.; Peijnenburg, A.; Stoopen, G.; Sharma, R.P.; Kumar, V.; Marx-Stoelting, P. Hepatotoxicity of the pesticides imazalil, thiacloprid and clothianidin–Individual and mixture effects in a 28-day study in female Wistar rats. *Food Chem. Toxicol.* **2020**, *140*, 111306. [CrossRef] [PubMed]

46. Abd-Elhakim, Y.M.; El Sharkawy, N.I.; El Bohy, K.M.; Hassan, M.A.; Gharib, H.S.; El-Metwally, A.E.; Arisha, A.H.; Imam, T.S. Iprodione and/or chlorpyrifos exposure induced testicular toxicity in adult rats by suppression of steroidogenic genes and SIRT1/TERT/PGC-1α pathway. *Environ. Sci. Pollut. Res.* **2021**, *28*, 56491–56506. [CrossRef] [PubMed]

Article

Quantitative Modeling of the Degradation of Pesticide Residues in Wheat Flour Supply Chain

Zhiqian Ding [1], Meirou Lin [1], Xuelin Song [2], Hua Wu [3] and Junsong Xiao [1,*]

[1] College of Food and Health, Beijing Technology & Business University (BTBU), Beijing 100048, China
[2] COFCO Grains Holding Co., Ltd., Beijing 100020, China
[3] College of Chemistry and Materials Engineering, Beijing Technology & Business University (BTBU), Beijing 100048, China
* Correspondence: xiaojs@btbu.edu.cn; Tel.: +86-134-3687-7820

Abstract: Pesticide residues in grain products are a major issue due to their comprehensive and long-term impact on human health, and quantitative modeling on the degradation of pesticide residues facilitate the prediction of pesticide residue level with time during storage. Herein, we tried to study the effect of temperature and relative humidity on the degradation profiles of five pesticides (carbendazim, bensulfuron methyl, triazophos, chlorpyrifos, and carbosulfan) in wheat and flour and establish quantitative models for prediction purpose. Positive samples were prepared by spraying the corresponding pesticide standards of certain concentrations. Then, these positive samples were stored at different combinations of temperatures (20 °C, 30 °C, 40 °C, 50 °C) and relative humidity (50%, 60%, 70%, 80%). Samples were collected at specific time points, ground, and the pesticide residues were extracted and purified by using QuEChERS method, and then quantified by using UPLC-MS/MS. Quantitative model of pesticide residues was constructed using Minitab 17 software. Results showed that high temperature and high relative humidity accelerate the degradation of the five pesticide residues, and their degradation profiles and half-lives over temperature and relative humidity varied among pesticides. The quantitative model for pesticide degradation in the whole process from wheat to flour was constructed, with R^2 above 0.817 for wheat and 0.796 for flour, respectively. The quantitative model allows the prediction of the pesticide residual level in the process from wheat to flour.

Keywords: wheat and flour; pesticide residues; storage conditions; ultra-high performance liquid chromatography–tandem mass spectrometry (UPLC-MS/MS); QuEChERS

Citation: Ding, Z.; Lin, M.; Song, X.; Wu, H.; Xiao, J. Quantitative Modeling of the Degradation of Pesticide Residues in Wheat Flour Supply Chain. *Foods* **2023**, *12*, 788. https://doi.org/10.3390/foods12040788

Academic Editor: Thierry Noguer

Received: 8 December 2022
Revised: 16 January 2023
Accepted: 6 February 2023
Published: 13 February 2023

1. Introduction

Wheat flour is one of the most common food ingredients in the world, and it occupies an important position in the food industry [1,2]. Pesticides are commonly used in wheat production and play a positive role in reducing crop yield losses caused by crop diseases, pests, and weeds [3]. The general population is exposed to pesticides primarily through eating food contaminated with pesticides residues. Although these pesticides are developed to function with minimal impact on human health, long-term exposure to pesticide residues in food remains a major risk. Controlling pesticide residues in the supply chains of food, especially rice and wheat flour, is important, and the ability to predict their degradation under various environmental conditions and processing factors is vital to their control [4]. Thus, in this research, the degradation profiles of commonly used pesticides in the supply chain of wheat flour, mainly including the storage and milling period of wheat flour, were studied.

The common pesticides in wheat include fungicides (e.g., carbendazim, chlorothanlonil, carboxin, cyproconazole, and difenoconazole), herbicides (e.g., bensulfuron methyl, carfentrazone-ethyl, dicamba, flupyrsulfuron methyl, and difenoconazole), insecticides (e.g., chlorpyrifos, triazophos, carbosulfan, deltamethrin, and esfenvalerate), plant growth regulators (e.g., ethephon, trinexapac, and chlormequat), of which five pesticides (carbendazim, bensulfuron methyl, triazophos, chlorpyrifos, and carbosulfan) are used more frequently in wheat cultivation. Pesticide residue problems often occur in wheat and wheat products [5–7].

The commonly used methods for pesticide extraction and purification include dispersive liquid–liquid microextraction, solid-phase extraction, solid-phase microextraction, QuEChERS (Quick, Easy, Cheap, Effective, Rugged, and Safe) method [8–11]. Among them, the QuEChERS method is one of the most successful methods [12], which is based on the principle of using absorbent filler to interact with impurities in the matrix to adsorb impurities and achieve the purpose of impurity removal. This method is fast, simple, economical, efficient, durable, and safe, which has been widely used in recent years for the study of pesticide residues in food [13].

The commonly used methods for the detection of pesticides in food are gas chromatography, gas chromatography–tandem mass spectrometry, liquid chromatography, liquid chromatography–tandem mass spectrometry, and ultra-high performance liquid chromatography–tandem mass spectrometry (UPLC-MS/MS) [14–17]. Gas chromatography is incapable of analyzing pesticides with high polarity and poor thermal stability [18]. Liquid chromatography–mass spectrometry is capable of detecting various pesticide residues, but its detection limit is often difficult to meet the requirements of pesticide residue detection [19]. Tandem mass technique, especially triple quadrupole mass spectrometry, coupled with ultra-high-performance liquid chromatography, is more comprehensive, sensitive, stable, and of wide detection range for the quantification of pesticide residues [20].

In this study, the degradation patterns of five typical pesticides during the storage of wheat and flour under various storage conditions were investigated, using QuEChERS method coupled with UPLC-MS/MS. The five pesticides included carbendazim, a broad-spectrum fungicide, bensulfuron methyl, an herbicide, triazophos, an organophosphorus acaricide, chlorpyrifos, a thiophosphate insecticide, and carbosulfan, a carbamate insecticide. A typical wheat supply chain includes the storage of wheat grain, the milling process, and the storage of wheat flour. The degradation profiles of these five pesticide residues, combined with their processing factors during wheat milling, will facilitate the model construction to predict the pesticide residue levels in the wheat supply chain.

2. Materials and Methods

2.1. Materials and Reagents

Wheat (Jimai 22) was provided by Crop Research Institute, Shandong Academy of Agricultural Sciences (Jinan, China). Flour (Fuqiang) was supplied by Beijing Guchuan Food Co., Ltd. (Beijing, China). The pesticide standards, including triphenyl phosphate (TPP), carbendazim, bensulfuron methyl, triazophos, chlorpyrifos, and carbosulfan (all 99% purity), were purchased from Accustandard Inc. (New Haven, CT, USA). The five pesticides (carbendazim, bensulfuron methyl, triazophos, chlorpyrifos, carbosulfan) chemical structures are shown in Figure 1. Chromatographic-grade methanol and formic acid were purchased from Mreda (Beijing, China), and 0.22 μm nylon membrane filter was purchased from Tianjin jinteng Experimental Equipment Co., Ltd. (Tianjin, China). Waters BEH C_{18} column (100 mm × 2.1 mm, 1.7 μm) was purchased from Waters (Milford, CT, USA). QuEChERS purifier and salt package were purchased from Beijing Dima Outai Science Technology Co., Ltd. (Beijing, China).

2.2. Standard Solutions

Stock solutions of each pesticide standard (100 mg/L) in the same volumetric flask were prepared with chromatographic grade methanol. The stock solutions were prepared with methanol as a mixed stock solution (2 mg/L) and stored at 4 °C in a refrigerator for use. A series of concentrations of working solutions (0.005, 0.020, 0.050, 0.200, 0.400, 1.000 mg/L) were obtained by sequentially diluting the mixed stock solution with wheat and flour blank matrix solution.

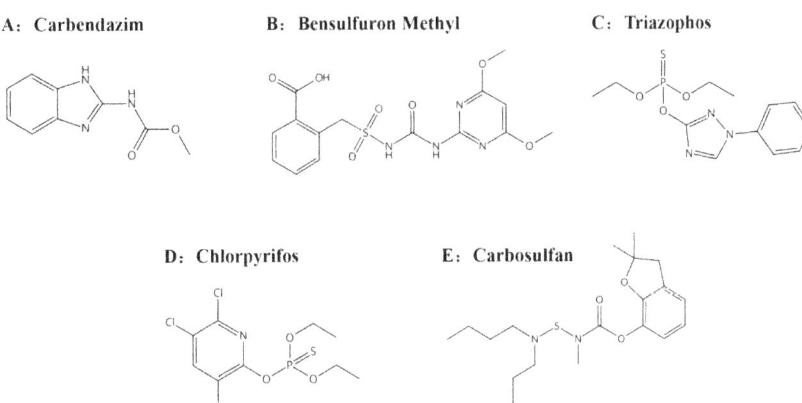

Figure 1. Chemical structures of five pesticides.

2.3. Preparation of Positive Samples

According to the standard "Maximum Residue Limits of Pesticides in Food of National Food Safety Standard" (GB 2763-2021), and the limit of detection (LOD) and limit of quantification (LOQ) of the experiment, a mixed solution containing 10-fold maximum residue limit (MRL) of carbendazim, triazophos, and carbosulfan, 1-fold MRL of chlorpyrifos, and 20-fold MRL of bensulfuron methyl were prepared in methanol. Then, 5.0 g of each wheat and flour samples were sprayed with the above prepared mixed solution and left sealed for 24 h for storage experiments.

2.4. Control of Sample Storage Conditions

Different concentrations of glycerol were prepared and placed into closed drying containers to obtain environments with relative humidity of 50%, 60%, 70%, 80%, respectively. The wheat and flour samples sprayed with pesticides were placed in containers with different humidity control, and then each container was placed in incubators with constant temperature at 20 °C, 30 °C, 40 °C, and 50 °C, respectively. Different combinations of temperature and humidity were obtained.

2.5. Extraction and Purification of Samples

The extraction and purification methods for wheat and flour were based on the classical QuEChERS method with minor change [21–23]. Wheat and flour samples were ground and weighed (5 ± 0.02 g) in a polypropylene centrifuge tube (50 mL), and TPP (100 µL) and pure water (10 mL) were added to mix, and the samples were left for 10 min to infiltrate. After immersion, samples were extracted by adding 10 mL methanol and salt packets (1.5 g CH_3COONa, 6 g anhydrous $MgSO_4$) and vortexed to mix [24–26]. After centrifugation at 4000× *g* for 10 min, the supernatant (6 mL) was collected in a centrifuge tube (10 mL). The samples were purified by adding a purifier (400 mg PSA, 400 mg C18, anhydrous $MgSO_4$) to the supernatant, and then vortexed and mixed. After centrifugation at 2000× *g* for 10 min, the supernatant (2 mL) was aspirated into a new centrifuge tube (10 mL). After blowing nitrogen to near-dryness at 40 °C, 2 mL methanol was added to redissolve, and then vortexed and mixed. Finally, the purified extracts were filtered through a 0.22 µm nylon membrane filter to analyze with UPLC-MS/MS.

2.6. Conditions for the UPLC-MS/MS Analysis

The five pesticides were separated on a UPLC-MS/MS (Waters ACQUITY UPLC I-Class/Xevo TQ-S) (Waters, Milford, CT, USA) equipped with positive mode (ESI+) and Waters BEH C_{18} column (100 mm × 2.1 mm, 1.7 µm). The mobile phase consisted of 0.1% formic acid in a mixed solvent of water (A) and acetonitrile (B). The gradient elution procedure was as follows: 10% B (0–1.5 min), 50% B (1.5–4 min), 90% B (4–10.5 min), and 10% B (10.5–13 min). The flow rate of the mobile phase was set at 0.4 mL/min, and the injection volume was 5 µL. The samples were measured by multiple reaction monitoring (MRM) in positive ion. The MRM parameters are shown in Table 1. The parameters of MS detection were as follows: capillary voltage, 3.0 KV; ion source temperature, 150 °C; desolvation temperature, 500 °C; desolvation air flow, 800 L/h; and cone voltage, 35 V. The quantitative ion chromatograms for the five pesticides are shown in Figure 2.

Table 1. Mass spectrometric parameters of five pesticides.

Pesticide Name	Retention Time/min	Qualitative Ion Pair	Quantitative Ion Pair	Tapered Hole Voltage (V)	Collision Energy (eV)
Carbendazim	3.38	192.1–132.1	192.1–160.1	18	35
Bensulfuron Methyl	6.65	411.2–91.14	411.2–182.1	27	20
Triazophos	7.19	314.1–118.9	314.1–161.9	18	35
Chlorpyrifos	8.58	349.9–97	349.9–198	18	35
Carbosulfan	9.36	381.2–118.2	381.2–160.1	18	35

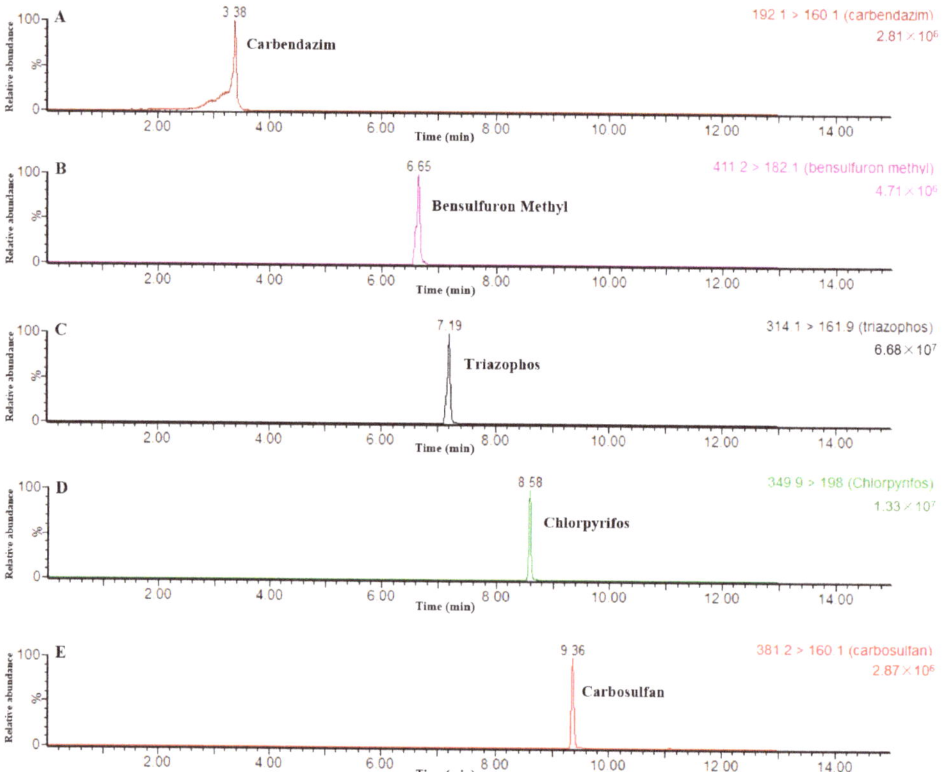

Figure 2. Chromatograms of quantitative ion pairs of five pesticides. Top to bottom: (**A**) Carbendazim, (**B**) Bensulfuron methyl, (**C**) Triazophos, (**D**) Chlorpyrifos, and (**E**) Carbosulfan.

2.7. Data Analysis

The data of this study were collected under Masslynx 4.1 software (Waters Corp., Milford, MA, USA). Microsoft Excel 2020 software (Microsoft Corp., Redmond, WA, USA) was used for preliminary sorting of experimental data, and Minitab 17 software (Minitab Inc., State College, PA, USA) was used for response surface analysis. Origin 2019b software

(OriginLab Corp., Northampton, MA, USA) was used to draw plots and calculate the Area Under Curve (AUC).

3. Results and Discussion

3.1. Standard Curves and Methodological Validation

In this study, the standard curves and coefficients of determination (R^2) of the five pesticides in wheat and flour blank matrices were obtained using the internal standard method. As shown in Tables 2 and 3, the R^2 of the standard curves of the five pesticides were all greater than 0.9990, indicating good linearity in a certain linear range.

Table 2. Standard working curves of five pesticides in wheat blank matrix solution.

Pesticides	Linear Range (µg/L)	Linear Equation	R^2
Carbendazim	5–2000	y = 0.1165x + 1.2126	0.9993
Bensulfuron Methyl	5–2000	y = 0.0460x − 0.0169	0.9999
Triazophos	5–2000	y = 0.3872x − 0.1041	0.9996
Chlorpyrifos	5–2000	y = 0.2607x − 0.2837	0.9995
Carbosulfan	20–2000	y = 0.0780x − 0.1619	0.9990

Table 3. Standard working curves of five pesticides in flour blank matrix solution.

Pesticides	Linear Range (µg/L)	Linear Equation	R^2
Carbendazim	5−1000	y = 0.1368x + 0.0838	0.9994
Bensulfuron Methyl	5−1000	y = 0.0347x + 0.0123	0.9995
Triazophos	5−1000	y = 0.1565x + 0.2937	0.9995
Chlorpyrifos	5−1000	y = 0.0634x + 0.0762	0.9993
Carbosulfan	5−1000	y = 0.0352x + 0.0020	0.9996

The LOD was obtained according to a three-fold Signal-to-Noise (S/N) ratio, and the LOQ was obtained according to a ten-fold S/N ratio [27]. Spike recoveries of the five pesticides were obtained by spiking the samples, wheat or flour, with a mixed standard solution of five pesticide residues at the level of 0.05 mg/kg. The experiment was set up in five parallel, and the results are shown in Table 4. The LOD and LOQ of five pesticide residues were 0.001~0.005 mg/kg and 0.002~0.01 mg/kg, respectively. The matrix spike recovery rate of pesticide residues was 86.77%~106.28%, and the precision was 2.88%~6.76%. It showed that the method has good recovery and precision and meets the requirements of pesticide residue quantification. This method can be used for the following experiments.

Table 4. The matrix spike recovery, relative standard deviation, limits of detection and limits of quantification of five pesticides.

Pesticides	Recovery (%) 0.05 mg/kg	Precision (%) 0.05 mg/kg	LOD (mg/kg)	LOQ (mg/kg)
Carbendazim	106.28%	3.36%	0.005	0.01
Bensulfuron Methyl	92.77%	5.32%	0.002	0.005
Triazophos	96.82%	6.62%	0.001	0.003
Chlorpyrifos	97.69%	2.88%	0.001	0.003
Carbosulfan	86.77%	6.76%	0.001	0.002

3.2. Critical Points of Pesticide Residue Change in Wheat Flour Supply Chain

The supply chain from wheat to flour includes raw material storage, cleaning, milling, sieving, cleaning, final product storage, packaging, and circulation, etc. Through previous literature research and investigation, we found that raw material storage, milling, and final product storage are the critical points in wheat flour supply chain [28,29]. The critical points and their main mechanisms and impact factors pesticide residues are shown in Table 5.

Table 5. Critical points in wheat flour supply chain.

Critical Points	Main Mechanisms	IMPACT FACTORS
Raw material storage	Pesticide residue degradation	Storage time (t_w), Temperature (T_w), Relative humidity (RH_w)
Milling	Physical removal of the cortex and Pesticide residue degradation	Mass fraction of pesticide residues in the cortex
Final product storage	Pesticide residue degradation	Storage time (t_f), Temperature (T_f), Relative humidity (RH_f)

Figure 3 showed the supply chain of wheat flour, from the storage of wheat grain, the milling process of wheat, to the storage of flour. Among them, C_0, C_1, C_2, and C_3 were pesticide residue concentrations at corresponding stage, and T_w, t_w, and RH_w are the temperature, time, and relative humidity during wheat storage, and T_f, t_f, and RH_f are the temperature, time, and relative humidity during storage. The processing factor (PF) was used to describe the effect of milling process on the degradation of pesticides degradation and defined as the ratio of C_2 to C_1. PF less than 1 indicates that the processing method can effectively reduce the amount of pesticide residues, and the lower the PF value, the lower the amount of pesticide residues [30,31].

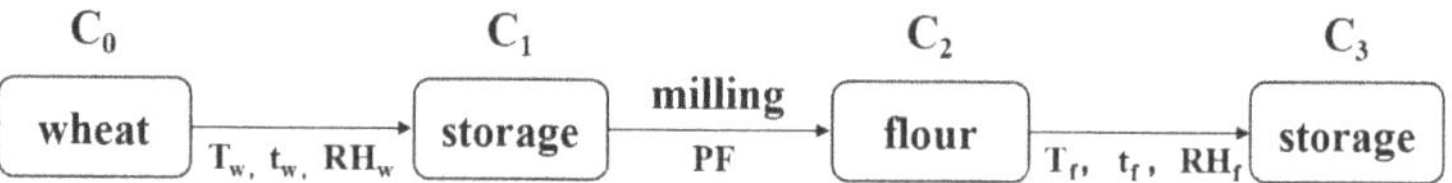

Figure 3. Critical processes in wheat flour processing.

The changes in pesticide residues in the wheat and flour storage process are complex, and it is susceptible to the influence of storage conditions such as temperature and relative humidity. Therefore, the degradation pattern of pesticide residues during wheat and flour storage is mainly explored in the following study to construct the degradation model of pesticide residues in the wheat flour supply chain.

3.3. Study on the Degradation Pattern of Five Pesticide Residues in Wheat and Flour

The second-order mathematical models of the degradation of five pesticides in wheat and flour at different times, temperatures, and relative humidity were constructed using Minitab 17 software, and the degradation models of five pesticides at different storage conditions are shown in Tables 6 and 7.

Table 6. A mathematical model for predicting the change of five pesticides during wheat storage.

Pesticide Name	Prediction Model	R^2
Carbendazim	$C_1/C_0 = 0.438 - 0.00246t_w + 0.0217T_w - 0.00184RH_w + 0.0000736t_w^2 - 0.000312T_w^2 - 0.00004015(RH_w)^2 + 0.0000674T_wt_w + 0.00000585T_wRH_w + 0.00005695T_wRH_w$	0.830
Bensulfuron Methyl	$C_1/C_0 = 0.5 - 0.034575t_w + 0.035875T_w - 0.007RH_w + 0.000285t_w^2 - 0.00047T_w^2 + 0.000035(RH_w)^2 + 0.000025T_wt_w - 0.00003t_wRH_w + 0.000001T_wRH_w$	0.817
Triazophos	$C_1/C_0 = 0.298 - 0.02986t_w + 0.046T_w - 0.00346RH_w + 0.000244t_w^2 - 0.000666T_w^2 - 0.000004(RH_w)^2 - 0.000008t_wT_w + 0.000002t_wRH_w + 0.000036T_wRH_w$	0.852

Table 6. *Cont.*

Pesticide Name	Prediction Model	R^2
Chlorpyrifos	$C_1/C_0 = 0.586 - 0.03286t_w + 0.02572T_w - 0.00434RH_w + 0.000276t_w{}^2 - 0.000328T_w{}^2 + 0.000016(RH_w)^2 - 0.000028t_wT_w + 0.00001t_wRH_w + 0.000006T_wRH_w$	0.868
Carbosulfan	$C_1/C_0 = 0.728 - 0.02008t_w + 0.01628T_w - 0.00171RH_w + 0.000157t_w{}^2 - 0.000259T_w{}^2 - 0.000007(RH_w)^2 + 0.000019t_wT_w - 0.000025t_wRH_w - 0.000017T_wRH_w$	0.863

Note: T_w, wheat storage temperature; t_w, wheat storage time; RH_w, wheat storage relative humidity; C_0, wheat initial pesticide residue concentration; C_1, wheat pesticide residue concentration after storage.

Table 7. A mathematical model for predicting the change of five pesticides during flour storage.

Pesticide Name	Prediction Model	R^2
Carbendazim	$C_3/C_2 = 1.756 - 0.05176t_f - 0.0042T_f - 0.01902RH_f + 0.000575t_f{}^2 + 0.000043T_f{}^2 + 0.0001(RH_f)^2 + 0.0000412t_fT_f + 0.0000768t_fRH_f - 0.0000162T_fRH_f$	0.796
Bensulfuron Methyl	$C_3/C_2 = 1.738 - 0.05116t_f + 0.0033T_f + 0.0185RH_f + 0.000549t_f{}^2 - 0.000095T_f{}^2 + 0.00007(RH_f)^2 + 0.000006T_ft_f + 0.000116t_fRH_f + 0.000081T_fRH_f$	0.802
Triazophos	$C_3/C_2 = 1.253 - 0.0411t_f - 0.0022T_f - 0.006RH_f + 0.000491t_f{}^2 + 0.00021T_f{}^2 - 0.0000165(RH_f)^2 + 0.0000065t_fT_f + 0.0000115t_fRH_f + 0.0001T_fRH_f$	0.878
Chlorpyrifos	$C_3/C_2 = 0.968 - 0.05488t_f - 0.0106T_f - 0.00196RH_f + 0.00063t_f{}^2 - 0.000192T_f{}^2 + 0.00002(RH_f)^2 + 0.000018t_fT_f + 0.0000656t_fRH_f + 0.000005T_fRH_f$	0.840
Carbosulfan	$C_3/C_2 = 1.314 - 0.03604t_f + 0.01218T_f - 0.01RH_f + 0.000467t_f{}^2 + 0.000225T_f{}^2 + 0.0000344(RH_f)^2 + 0.000023t_fT_f - 0.000045t_fRH_f - 0.000008T_fRH_f$	0.863

Note: T_f, flour storage temperature; t_f, flour storage time; RH_f, flour storage relative humidity; C_2, flour initial pesticide residue concentration; C_3, flour pesticide residue concentration after storage.

As can be seen in Tables 6 and 7, the R^2 of five pesticides is above 0.796 by building models, which indicates that the fitting result is good and reaches the expected level. By constructing the mathematical model, the degradation pattern of pesticide residues in wheat and flour under different storage conditions can be reasonably predicted.

3.4. Effect of Storage Conditions on the Degradation of Five Pesticides in Wheat and Flour

3.4.1. Effect of Storage Temperature on the Degradation of Pesticide Residues

The degradation patterns of pesticide residues of carbendazim, bensulfuron methyl, triazophos, chlorpyrifos, and carbosulfan in wheat and flour at different storage temperatures are shown in Figures 4 and 5.

As shown in Figure 4, the five pesticide residues decreased with storage time during the 90-day storage period of wheat. Under the four temperatures from low to high, the carbendazim residue decreased by 86.83%, 89.47%, 91.63%, and 98.14% with the half-lives of 10.27, 7.33, 8.04, and 6.97 days, respectively, the bensulfuron methyl residue decreased by 96.00%, 97.63%, 98.31%, and 99.00% with the half-lives of 8.93, 9.11, 7.50, and 5.37 days respectively, the triazophos residue decreased by 90.87%, 91.05%, 98.30%, and 98.80% with the half-lives of 13.47, 11.16, 8.48, and 6.61 days, respectively, the chlorpyrifos residue decreased by 97.40%, 97.85%, 98.45%, 98.70% with the half-lives of 10.09, 9.64, 9.11, and 6.43 days, respectively, the carbosulfan residue decreased by 77.85%, 84.33%, 84.93%, and 87.95% with half-lives of 20.42, 11.87, 12.67, and 9.73 days, respectively. In conclusion, the degradation rates of the five pesticides increased with increasing temperature and reached a peak at 50 °C.

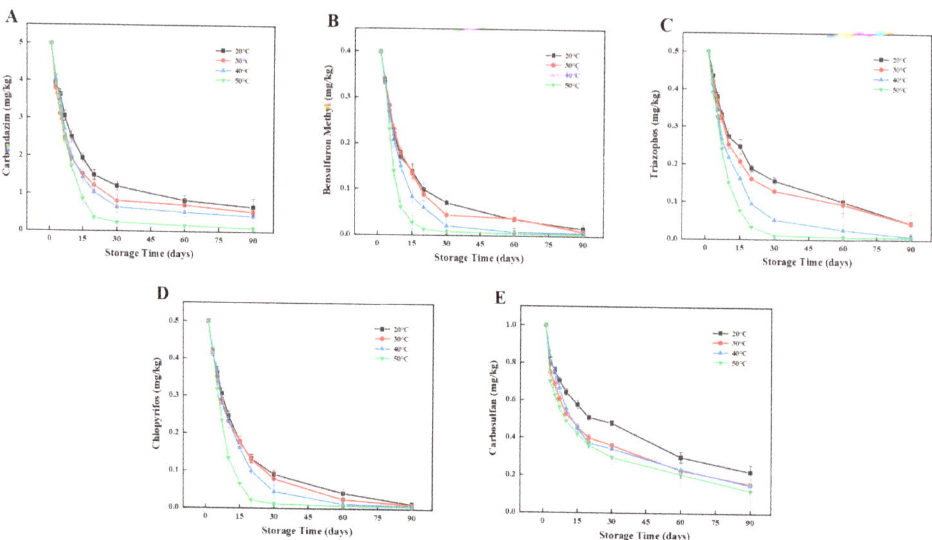

Figure 4. Effects of temperature on degradation of five pesticides during storage of wheat. Left to right: (**A**) Carbendazim, (**B**) Bensulfuron methyl, (**C**) Triazophos, (**D**) Chlorpyrifos, and (**E**) Carbosulfan.

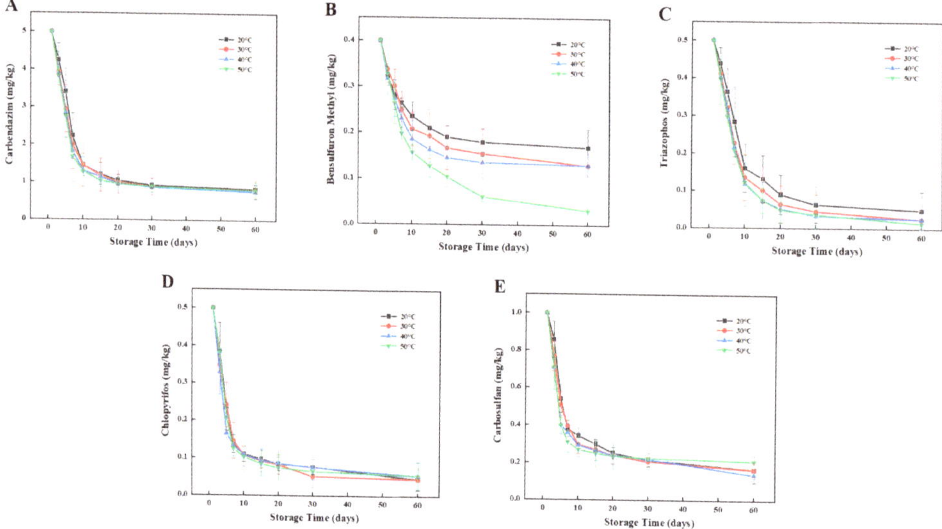

Figure 5. Effects of temperature on the degradation of five pesticides during storage of flour. Left to right: (**A**) Carbendazim, (**B**) Bensulfuron methyl, (**C**) Triazophos, (**D**) Chlorpyrifos, and (**F**) Carbosulfan.

As shown in Figure 5, the five pesticide residues decreased with storage time during the 60-day storage period of flour. Under the four temperatures from low to high, the carbendazim residue decreased by 83.41%, 83.83%, 84.74%, and 84.00% with the half-lives of 6.73, 5.90, 5.55, and 5.25 days, respectively, the bensulfuron methyl residue decreased by 57.74%, 67.54%, 67.36%, and 92.25% with the half-lives of 15.23, 12.04, 9.15, and 7.91 days, respectively, the triazophos residue decreased by 90.22%, 94.92%, 94.94%, and 95.21% with the half-lives of 7.85, 6.49, 6.02, and 5.84 days, respectively, the chlorpyrifos residue decreased by 82.40%, 83.81%, 90.87%, and 98.62% with the half-lives of 4.66, 4.60, 3.78, and 4.37 days, respectively, the carbosulfan residue decreased by 83.48%, 83.58%, 86.38%, and

79.18% with half-lives of 6.08, 5.49, 4.54, and 4.25 days, respectively. In conclusion, the degradation rates of the five pesticides increased with increasing temperature and reached a peak at 50 °C.

In general, with the increase of storage temperature, the chemical reaction rate of pesticide degradation process is accelerated, and the volatility of the pesticide is enhanced, so the half-lives of all pesticide residues tested decreased. However, the susceptibility to temperature of the five pesticides are different, with carbosulfan as the most unsusceptible one. Temperature influences the volatility of pesticides, and those with lower boiling point may be more susceptible. Temperature also promotes the chemical reaction rate involved in the process of pesticide degradation, and Q_{10} coefficient of these reactions to temperature are different, and this may also explain the different susceptibility to temperature of pesticides [32].

3.4.2. Effect of Relative Humidity during Storage on the Degradation of Pesticide Residues

The degradation patterns of pesticide residues of carbendazim, bensulfuron methyl, triazophos, chlorpyrifos, and carbosulfan in wheat and flour at different storage relative humidity are shown in Figures 6 and 7.

As shown in Figure 6, the five pesticide residues decreased with storage time during the 90-day storage period of wheat. Under the four relative humidity from low to high, the carbendazim residue decreased by 89.14%, 90.57%, 91.72%, and 94.64% with the half-lives of 8.93, 8.31, 7.68, and 7.15 days, respectively, the bensulfuron methyl residue decreased by 96.50%, 97.88%, 98.44%, and 98.75% with the half-lives of 8.39, 7.68, 7.06, and 6.61 days, respectively, the triazophos residue decreased by 90.20%, 93.50%, 96.65%, and 98.60% with the half-lives of 10.62, 9.64, 8.75, and 8.04 days, respectively, the chlorpyrifos residue decreased by 96.60%, 98.07%, 98.33%, and 98.60% with the half-lives of 10.62, 9.91, 9.29, and 8.66 days, respectively, the carbosulfan residue decreased by 77.83%, 83.00%, 85.00%, and 89.23% with half-lives of 14.01, 13.70, 12.76, and 12.05 days, respectively. In conclusion, the degradation rates of the five pesticides increased with increasing relative humidity and reached a peak at 80%.

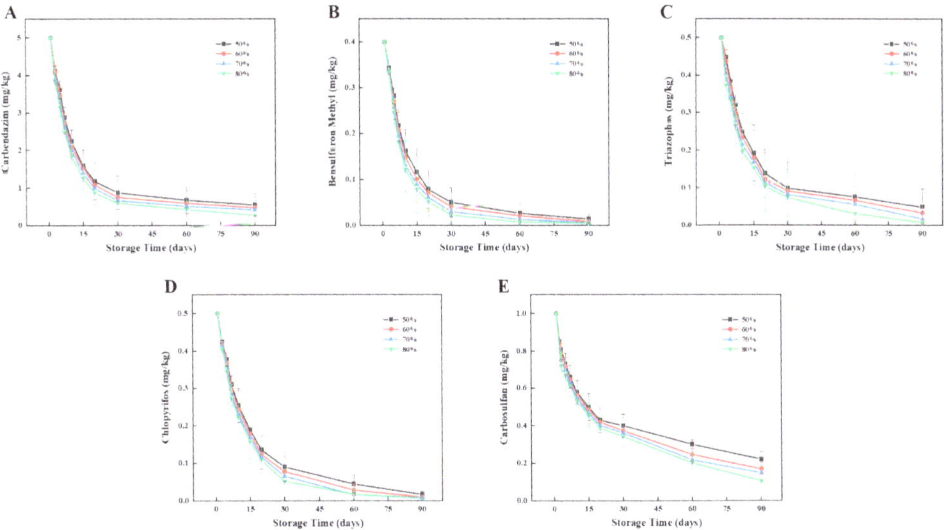

Figure 6. Effects of relative humidity on degradation of five pesticides during storage of wheat. Left to right: (**A**) Carbendazim, (**B**) Bensulfuron methyl, (**C**) Triazophos, (**D**) Chlorpyrifos, and (**E**) Carbosulfan.

As shown in Figure 7, the five pesticide residues decreased with storage time during the 60-day storage period of flour. Under the four relative humidity states from low to high, the carbendazim residue decreased by 79.84%, 82.46%, 86.96%, and 86.72% with the half-lives of 19.43, 11.75, 7.56, and 6.02 days, respectively, the bensulfuron methyl residue decreased by 60.32%, 63.98%, 70.75%, and 73.35% with the half-lives of 8.91, 7.56, 5.78, and 4.84 days respectively, the triazophos residue decreased by 84.91%, 90.94%, 98.70%, and 99.24% with the half-lives of 5.55, 4.51, 4.07, and 3.60 days, respectively, the chlorpyrifos residue decreased by 79.62%, 79.79%, 88.59%, and 89.42% with the half-lives of 6.20, 5.40, 4.78, and 4.19 days, respectively, the carbosulfan residue decreased by 90.79%, 90.89%, 88.88%, and 88.92% with half-lives of 6.08, 5.49, 4.54, and 4.25 days, respectively. In conclusion, the degradation rates of the five pesticides increased with increasing relative humidity and reached a peak at 80%.

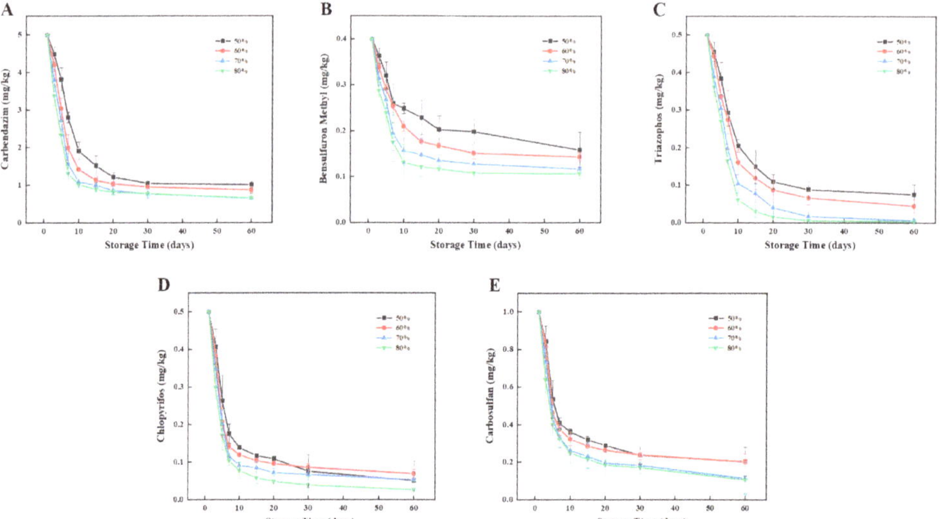

Figure 7. Effects of relative humidity on degradation of five pesticides during storage of flour. Left to right: (**A**) Carbendazim, (**B**) Bensulfuron methyl, (**C**) Triazophos, (**D**) Chlorpyrifos, and (**E**) Carbosulfan.

In general, with the increase of storage relative humidity, the chemical reaction rate of the pesticide degradation process is accelerated, and the half-lives of all pesticide residues tested decreased. However, different structures of pesticides have different sensitivity to relative humidity. Relative humidity influences the hydrolysis reaction of pesticides, and those with more ester bonds may be more susceptible [33]. Relative humidity also promotes the growth and reproduction rate of microorganisms involved in the process of pesticide degradation, and microbial activity has an effect on the degradation process of pesticides, and this could also explain the different susceptibility to relative humidity of pesticides [34].

3.4.3. Interaction between Storage Temperature and Relative Humidity on Pesticide Residue Degradation

The interaction between temperature and relative humidity on degradation of pesticide residues was studied by response surface analysis using Minitab 17 software, as shown in Figures 8 and 9. AUC is the area integral under the pesticide residue degradation curve for a specific combination of temperature and relative humidity, which reflects a comprehensive measure of the level of pesticide degradation with time at a specific temperature and relative humidity. The increase of AUC indicates the decrease of pesticide degradation rate, and the decrease of AUC indicates the increase of pesticide degradation rate. The interaction relationship between temperature and relative humidity on AUC of wheat stored for 90 days is shown in Figure 8, with four temperatures at 20 °C, 30 °C, 40 °C, 50 °C, and four relative humidity at 50%, 60%, 70%, 80%, respectively. The analysis showed that both temperature and relative humidity significantly ($p < 0.05$) affected the AUC of the five pesticides. With the increases of temperature and relative humidity, the AUC of five pesticides increased to a certain extent and then decreased.

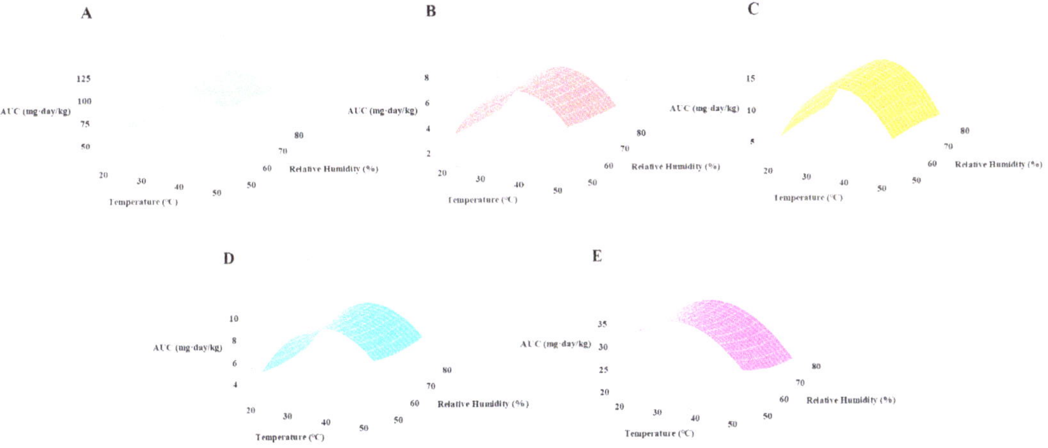

Figure 8. Effects of temperature and relative humidity on degradation of five pesticides during storage of wheat. Left to right: (**A**) Carbendazim, (**B**) Bensulfuron methyl, (**C**) Triazophos, (**D**) Chlorpyrifos, and (**E**) Carbosulfan.

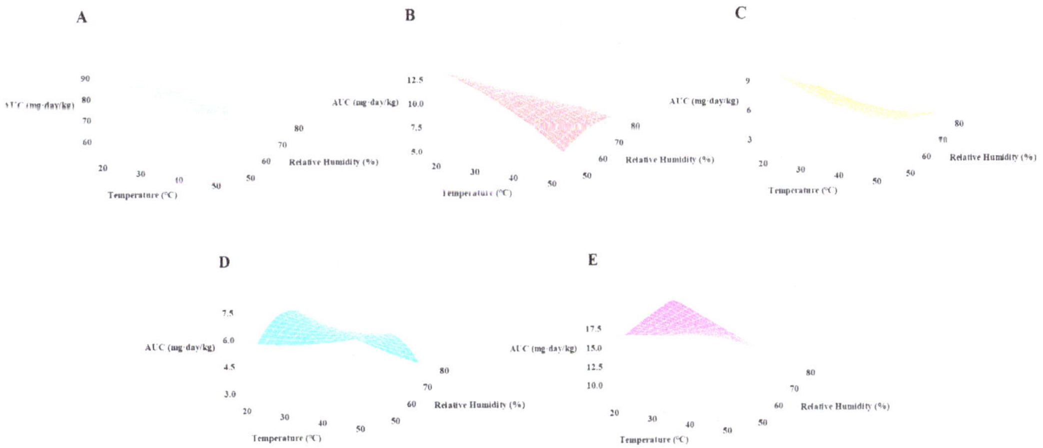

Figure 9. Effects of temperature and relative humidity on degradation of five pesticides during storage of flour. Left to right: (**A**) Carbendazim, (**B**) Bensulfuron methyl, (**C**) Triazophos, (**D**) Chlorpyrifos, and (**E**) Carbosulfan.

The interactional relationship between temperature and relative humidity on AUC of flour stored for 60 days is shown in Figure 9, with four temperatures at 20 °C, 30 °C, 40 °C, and 50 °C, and four relative humidity at 50%, 60%, 70%, and 80%, respectively. The results showed that temperature and relative humidity could significantly ($p < 0.05$) affect the AUC of carbendazim, bensulfuron methyl, and triazophos, and relative humidity could significantly ($p < 0.05$) affect the AUC of chlorpyrifos and carbosulfan. There was a significant ($p < 0.05$) interaction between temperature and relative humidity on the AUC of triazophos, chlorpyrifos, and carbosulfan, with p values of 0.031, 0.032, and 0.038, respectively. The AUC of five pesticides decreased with increasing temperature and relative humidity.

The reason may be that different structure of pesticides have different sensitivity to temperature and relative humidity. In general, high temperatures make the molecular structure of pesticides vulnerable to damage, and the degradation rate of the pesticide increases with the temperature. Changes of relative humidity can affect the hydrolysis mechanism of pesticides due to interaction between lipid molecules in pesticides and water molecules, destroying the structure of pesticide molecules and leading to the degradation of pesticide residues. Therefore, pesticides with different structures have different rates of residue degradation [35].

4. Conclusions

In this study, the five pesticides in wheat and flour were extracted and purified by the QuEChERS method and quantified by UPLC-MS/MS. The linear range, linear equation, LOD, LOQ, recovery rate, and precision of the method were investigated, and the results showed that the method was simple, rapid, with high accuracy and good applicability.

A quantitative model was constructed to predict the pesticide residue degradation during the storage of wheat and flour. The results showed that the R^2 reached above 0.817 in wheat, and the R^2 reached above 0.796 in flour, with good fitting effect. The model could be used to predict the degradation of pesticide residues at given time points of the wheat flour supply chain from wheat grain to the final product, i.e., wheat flour.

In flour and wheat, the five pesticide residues gradually decreased with the increase of storage time, and the degradation rate was faster in the early stage and slower in the later stage. The degradation rates of the five pesticides increased with increasing temperature and reached a peak at 50 °C. The degradation rates of the five pesticides increased with increasing relative humidity and reached a peak at 80%. Temperature may influence the volatility of pesticides and the rate of chemical reactions involved in the degradation process. Relative humidity may influence the hydrolysis reaction of pesticides and the growth and reproduction rate of microorganisms. Results showed that high temperature and high relative humidity accelerate the degradation of the five pesticides residues, and their degradation profiles and half-lives over temperature and relative humidity varied among pesticides.

Author Contributions: Conceptualization, J.X.; data curation, Z.D. and M.L.; investigation, Z.D. and M.L.; methodology, Z.D. and M.L.; project administration, H.W.; resources, J.X., H.W. and X.S.; software, Z.D.; supervision, J.X. and H.W.; visualization, Z.D., M.L. and J.X.; writing—original draft, Z.D.; writing—review and editing, Z.D., M.L. and J.X. All authors have read and agreed to the published version of the manuscript.

Funding: This research was funded by the National Key Research and Development Program of China (No. 2017YFC1600605).

Institutional Review Board Statement: Not applicable.

Informed Consent Statement: Not applicable.

Data Availability Statement: The data presented in this study are available on request from the corresponding author.

Conflicts of Interest: Author Xuelin Song was employed by the company COFCO Grains Holding Co., Ltd. The remaining authors declare that the research was conducted in the absence of any commercial or financial relationships that could be construed as a potential conflict of interest.

References

1. Li, M.; Zhu, K.X.; Guo, X.N.; Brijs, K.; Zhou, H.M. Natural additives in wheat-based pasta and noodle products: Opportunities for enhanced nutritional and functional properties. *Compr. Rev. Food Sci. Food Saf.* **2014**, *13*, 347–357. [CrossRef] [PubMed]
2. Twinomuhwezi, H.; Awuchi, C.G.; Rachael, M. Comparative study of the proximate composition and functional properties of composite flours of amaranth, rice, millet, and soybean. *Am. J. Food Sci. Nutr.* **2020**, *6*, 6–19.
3. Gomes, H.D.O.; Menezes, J.M.C.; da Costa, J.G.M.; Coutinho, H.D.M.; Teixeira, R.N.P.; do Nascimento, R.F. A socio-environmental perspective on pesticide use and food production. *Ecotoxicol. Environ. Saf.* **2020**, *197*, 110627. [CrossRef]
4. Grewal, A.S. Pesticide Residues in Food Grains, Vegetables and Fruits: A Hazard to Human Health. *J. Med. Chem. Toxicol.* **2017**, *2*, 40–46. [CrossRef]
5. Medina, M.B.; Munitz, M.S.; Resnik, S.L. Fate and health risks assessment of some pesticides residues during industrial rice processing in Argentina. *J. Food Compos. Anal.* **2021**, *98*, 103823. [CrossRef]
6. Mahugija, J.A.M.; Kayombo, A.; Peter, R. Pesticide residues in raw and processed maize grains and flour from selected areas in Dar es Salaam and Ruvuma, Tanzania. *Chemosphere* **2017**, *185*, 137–144. [CrossRef]
7. Medina, M.B.; Munitz, M.S.; Resnik, S.L. Pesticides in randomly collected rice commercialised in Entre Ríos, Argentina. *Food Addit. Contam.* **2019**, *12*, 252–258. [CrossRef]
8. Wang, X.C.; Shu, B.; Li, S.; Yang, Z.G.; Qiu, B. QuEChERS followed by dispersive liquid–liquid microextraction based on solidification of floating organic droplet method for organochlorine pesticides analysis in fish. *Talanta* **2017**, *162*, 90–97. [CrossRef]
9. Günter, A.; Balsaa, P.; Werres, F.; Schmidt, T.C. Influence of the drying step within disk-based solid-phase extraction both on the recovery and the limit of quantification of organochlorine pesticides in surface waters including suspended particulate matter. *J. Chromatogr. A* **2016**, *1450*, 1–8. [CrossRef]
10. Liang, W.; Wang, J.; Zang, X.; Dong, W.; Wang, C.; Wang, Z. Barley husk carbon as the fiber coating for the solid-phase microextraction of twelve pesticides in vegetables prior to gas chromatography–mass spectrometric detection. *J. Chromatogr. A* **2017**, *1491*, 9–15. [CrossRef]
11. Xu, X.; Zhang, X.; Duhoranimana, E.; Zhang, Y.; Shu, P. Determination of methenamine residues in edible animal tissues by HPLC-MS/MS using a modified QuEChERS method: Validation and pilot survey in actual samples. *Food Control.* **2016**, *61*, 99–104. [CrossRef]
12. Lehotay, S.J.; Son, K.A.; Kwon, H.; Koesukwiwat, U.; Fu, W.; Mastovska, K.; Hoh, E.; Leepipatpiboon, N. Comparison of QuEChERS sample preparation methods for the analysis of pesticide residues in fruits and vegetables. *J. Chromatogr. A* **2010**, *1217*, 2548–2560. [CrossRef] [PubMed]
13. Hakami, R.A.; Aqel, A.; Ghfar, A.A.; ALOthman, Z.A.; Badjah-Hadj-Ahmed, A.Y. Development of QuEChERS extraction method for the determination of pesticide residues in cereals using DART-ToF-MS and GC-MS techniques. Correlation and quantification study. *J. Food Compos. Anal.* **2021**, *98*, 103822. [CrossRef]
14. Jiang, J.; Li, P.W.; Xie, L.H.; Ding, X.X.; Li, Y.; Wang, X.P.; Wang, X.F.; Zhang, Q.; Guan, D. Rapid Screening and Simultaneous Confirmation of 64 Pesticide Residues in Vegetable Using Solid Phase Extraction-comprehensive Two-dimensional Gas Chromatography Coupled with Time-of-Flight Mass Spectrometer. *Chin. J. Anal. Chem.* **2011**, *39*, 72–76. [CrossRef]
15. Banerjee, T.; Banerjee, D.; Roy, S.; Banerjee, H.; Pal, S. A comparative study on the persistence of imidacloprid and beta-cyfluthrin in vegetables. *Bull. Environ. Contam. Toxicol.* **2012**, *89*, 193–196. [CrossRef]
16. Van der Heeft, E.; Bolck, Y.J.C.; Beumer, B.; Nijrolder, A.W.J.M.; Stolker, A.A.M.; Nielen, M.W.F. Full-scan accurate mass selectivity of ultra-performance liquid chromatography combined with time-of-flight and orbitrap mass spectrometry in hormone and veterinary drug residue analysis. *J. Am. Soc. Mass Spectrom.* **2011**, *20*, 451–463. [CrossRef]
17. Walorczyk, S.; Drożdżyński, D.; Kierzek, R. Determination of pesticide residues in samples of green minor crops by gas chromatography and ultra performance liquid chromatography coupled to tandem quadrupole mass spectrometry. *Talanta* **2015**, *132*, 197–204. [CrossRef]
18. Yao, X.L.; Han, D.; Liu, X.D.; Luo, Y.; Wu, Q.; An, H.M.; Wu, X.M. Simultaneous Determination of 11 Pesticides in Rosa roxburghii by QuEChERS-Ultra-high Performance Liquid Chromatography-Tandem Mass Spectrometry. *Food Sci.* **2022**, *43*, 309–316.
19. Li, J.F.; Jing, W.W.; Li, D.Q.; Pan, L. Determination of 7 Pesticide Residues in Xinjiang Grown Apple by Dispersive Solid-Phase Extraction Combined with Ultra Performance Liquid Chromatography-High Resolution Tandem Mass Spectrometry. *Food Sci.* **2018**, *39*, 295–301.
20. Xiu-Ping, Z.; Lin, M.; Lan-Qi, H.; Jian-Bo, C.; Li, Z. The optimization and establishment of QuEChERS-UPLC–MS/MS method for simultaneously detecting various kinds of pesticides residues in fruits and vegetables. *J. Chromatogr. B* **2017**, *1060*, 281–290. [CrossRef]
21. Lin, T.; Shao, J.L.; Liu, X.Y.; Yang, D.S.; Chen, X.L.; Li, Y.G.; Fan, J.L.; Liu, H.C. Determination of 41 pesticide residues in vegetables by QuEChERs-ultra perfoemance liquid chromatography-tandem mass spectrometry. *Chin. J. Chromatogr.* **2015**, *33*, 235–241. [CrossRef]
22. Cai, L.; Xi, P.Y.; Xei, Q.; Li, X.H.; Qiao, X.L.; Xie, H.B.; Chen, J.W.; Cai, X.Y. Q Development of a multiresidue method for determination of 110 pesticide residues in soil using QuEChERS-HPLC-MS/MS and QuEChERS-GC-MS. *J. Agro-Environ. Sci.* **2017**, *36*, 1680–1688.
23. Li, J.X.; Wang, Y.Z.; Lu, J.; Xu, D.; Shan, J.H.; Fan, B. Study on analysis of organophosphorus pesticides residue in honeysuckle and migration model during brewing process. *J. Food Sci. Technol.* **2020**, *38*, 119–126.

24. He, Z.; Wang, Y.; Xu, Y.; Liu, X. Determination of antibiotics in vegetables using QuEChERS-based method and liquid chromatography-quadrupole linear ion trap mass spectrometry. *Food Anal. Methods* **2018**, *11*, 2857–2864. [CrossRef]
25. Muhammad, N.; Subhani, Q.; Wang, F.; Lou, C.; Liu, J.; Zhu, Y. Simultaneous determination of two plant growth regulators in ten food samples using ion chromatography combined with QuEChERS extraction method (IC-QuEChERS) and coupled with fluorescence detector. *Food Chem.* **2018**, *241*, 308–316. [CrossRef]
26. Ajibola, A.S.; Tisler, S.; Zwiener, C. Simultaneous determination of multiclass antibiotics in sewage sludge based on QuEChERS extraction and liquid chromatography-tandem mass spectrometry. *Anal. Methods* **2020**, *12*, 576–586. [CrossRef]
27. Cui, L.L.; Yan, M.X.; Pang, S.F.; Zhou, C.Y.; Wang, Y.P. Simultaneous Determination of Seventeen Pesticide Residues in Ganoderma tsugae by Modified QuEChERS Combined with Gas Chromatography-Mass Spectrometry. *Food Sci.* **2019**, *40*, 326–331.
28. Thakur, S.; Scanlon, M.G.; Tyler, R.T.; Milani, A.; Paliwal, J. Pulse flour characteristics from a wheat flour miller's perspective: A comprehensive review. *Compr. Rev. Food Sci. Food Saf.* **2019**, *18*, 775–797. [CrossRef]
29. Dos Santos, J.L.P.; Bernardi, A.O.; Morassi, L.L.P.; Silva, B.S.; Copetti, M.V.; Sant'Ana, A.S. Incidence, populations and diversity of fungi from raw materials, final products and air of processing environment of multigrain whole meal bread. *Food Res. Int.* **2016**, *87*, 103–108. [CrossRef]
30. Zhang, Z.; Jiang, W.W.; Jian, Q.; Song, W.; Zheng, Z.; Wang, D.; Liu, X. Changes of field incurred chlorpyrifos and its toxic metabolite residues in rice during food processing from-RAC-to-consumption. *PLoS ONE* **2015**, *10*, e0116467. [CrossRef]
31. Liu, Y.; Su, X.; Jian, Q.; Chen, W.; Sun, D.; Gong, L.; Jiang, L.; Jiao, B. Behaviour of spirotetramat residues and its four metabolites in citrus marmalade during home processing. *Food Addit. Contam.* **2016**, *33*, 452–459. [CrossRef] [PubMed]
32. Liu, Y.; Qin, X.; Chen, Q.; Zhang, Q.; Yin, P.; Guo, Y. Effects of moisture and temperature on pesticide stability in corn flour. *J. Serb. Chem. Soc.* **2020**, *85*, 191–201. [CrossRef]
33. Farha, W.; El-Aty, A.M.A.; Rahman, M.; Shin, H.-C.; Shim, J.-H. An overview on common aspects influencing the dissipation pattern of pesticides: A review. *Environ. Monit. Assess.* **2016**, *188*, 693. [CrossRef]
34. Gatto, M.P.; Cabella, R.; Gherardi, M. Climate change: The potential impact on occupational exposure to pesticides. *Ann. Dell'istituto Super. Di Sanita* **2016**, *52*, 374–385.
35. Rubira, R.J.G.; Constantino, C.J.L.; Otero, J.C.; Sanchez-Cortes, S. Abiotic degradation of s-triazine pesticides analyzed by surface-enhanced Raman scattering. *J. Raman Spectrosc.* **2020**, *51*, 264–273. [CrossRef]

 *foods*

Article

Dissipation Kinetics and Risk Assessment of Spirodiclofen and Tebufenpyrad in *Aster scaber* Thunb

Ramesh Kumar Saini [1], Yongho Shin [2], Rakdo Ko [3], Jinchan Kim [3], Kwanghun Lee [3], Dai An [3], Hee-Ra Chang [4] and Ji-Ho Lee [1,*]

[1] Department of Crop Sciences, Konkuk University, Seoul 05029, Republic of Korea
[2] Department of Applied Biology, Dong-A University, Busan 49315, Republic of Korea
[3] Bio Division, Korea Conformity Laboratories, Incheon 21999, Republic of Korea
[4] Department of Food & Pharmaceutical Engineering, Graduate School of Hoseo University, Asan 31499, Republic of Korea
* Correspondence: micai1@naver.com; Tel.: +82-2-450-3758; Fax: +82-2-450-3754

Abstract: The dissipation kinetics of spirodiclofen and tebufenpyrad after their application on *Aster scaber* Thunb were studied for 10 days, including the pre-harvest intervals. Spirodiclofen and tebufenpyrad were used in two greenhouses in Taean-gun, Chungcheongnam province (Field 1) and Gwangyang-si, Jeollanam province (Field 2), Republic of Korea. Samples were taken at 0, 1, 3, 5, 7, and 10 days after pesticide application. The method validations were performed utilizing liquid chromatography (LC)-tandem mass spectrometry (MS/MS). The recoveries of the studied pesticides ranged from 82.0–115.9%. The biological half-lives of spirodiclofen and tebufenpyrad were 4.4 and 3.8 days in Field 1, and 4.5 and 4.2 days in Field 2, respectively. The pre-harvest residue limits (PHRLs; 10 days before harvesting) of *Aster scaber* were 37.6 mg/kg (Field 1) and 41.2 mg/kg (Field 2) for spirodiclofen, whereas the PHRLs were 7.2 (Field 1) and 3.6 (Field 2) for tebufenpyrad. The hazard quotient for both pesticides at pre-harvest intervals was less than 100% except in the case of spirodiclofen (0 day).

Keywords: pre-harvest residue limits (PHRLs); hazard quotient; *Doellingeria scabra* (Thunb.) Nees; pesticide; maximum residue limit; multiple reaction monitoring (MRM)

Citation: Saini, R.K.; Shin, Y.; Ko, R.; Kim, J.; Lee, K.; An, D.; Chang, H.-R.; Lee, J.-H. Dissipation Kinetics and Risk Assessment of Spirodiclofen and Tebufenpyrad in *Aster scaber* Thunb *Foods* **2023**, *12*, 242. https://doi.org/10.3390/foods12020242

Academic Editors: Guangyang Liu and Fengnian Zhao

Received: 10 November 2022
Revised: 28 December 2022
Accepted: 3 January 2023
Published: 5 January 2023

1. Introduction

Pesticides are widely used to control pests on crops effectively. The use of pesticides has resulted in increased productivity of cultivated crops. Thus, pesticides have become an economically essential agricultural resource for modern agriculture [1–3]. Despite the necessity for pesticides, the risks related to pesticide use and their residues in humans and the environment are being reported [4,5]. Therefore, considering the risks to consumers, the environment, and the productivity of crops, agricultural chemical management safety policies are being implemented nationwide. Pesticide residues in domestically distributed and imported agri-foods are being monitored in an effort to decrease the marketing and consumption of crops that exceed the maximum residue limit (MRL). Moreover, the Ministry of Food and Drug Safety has established and applied the pre-harvest residue limit (PHRL) in the production stage to systematize the management of pesticide residues before shipment and on a broader scale, to minimize damage to producers and consumers due to the distribution of substandard crops. According to the recommendations on the safe use of pesticides, PHRL is calculated and established using the biological half-life and decay constant, which are calculated each day before shipment after spraying the chemical [6]. Therefore, to provide safe food to consumers by preventing substandard crops from exceeding pesticide residue limits at the domestic distribution stage and to produce agricultural products meeting the pesticide residue tolerance limit of the exporting country, the establishment and management of PHRL is crucial [6].

Spirodiclofen, 3-(2,4-dichlorophenyl)-2-oxo-1-oxaspiro [4.5]dec-3-en-4-yl 2,2-dimethylbutyrate (Table A1) is an insecticide belonging to the chemical class of ketoenols or tetronic acids [7]. Spirodiclofen inhibits lipid biosynthesis and is approved for use on citrus, grapes, pears, nuts, and other plants in many countries to control pests and mites [7,8]. Tebufenpyrad is an electron transport chain inhibitor and a pyrazole acaricide that effectively inhibits *Tetranychus*, *Panonychus*, *Origonychus*, and *Eotetranychus* species [9].

Aster scaber Thunb. (Syn. *Doellingeria scabra* (Thunb.) Nees) is a perennial herb of the Asteraceae family, widely cultivated in the temperate region of Korea for culinary uses. It is a rich source of vitamins, minerals, and essential amino acids, which help minimize the incidence of chronic diseases [10]. It also contains a large amount of unsaturated fatty acids, such as linolenic acid, which lowers blood cholesterol [11].

Dissipation kinetics is used to determine the in-plant biological half-lives and pre-harvest intervals (PHIs). Therefore, in this study, the basic data for establishing the pesticide residue tolerance standards in the production stage were obtained by measuring the residual amount each day during cultivation after spraying *Aster scaber* with spirodiclofen and tebufenpyrad, which are used to eliminate *Tetranychus urticae* Koch and *Tetranychus kanzawai*. Through a risk assessment during the PHI period, the study sought to identify the risks of ingesting *Aster scaber* foods sprayed with spirodiclofen and tebufenpyrad.

2. Materials and Methods

2.1. Test Chemicals and Reagents

The pesticide standards, spirodiclofen and tebufenpyrad, were purchased from Kemidas, Gunpo-si, Gyeonggi-do, Republic of Korea. The chemicals spirodiclofen (36% wettable powder, Bayer CropScience, Seoul, Republic of Korea) and tebufenpyrad (10% emulsifiable concentrate, Syngenta, Seoul, Republic of Korea) were purchased as commercial products. The chemical structures and physicochemical properties of the two pesticides are shown in Table A1 [12]. An HPLC grade solvent was purchased from J.T. Baker® (Avantor Performance Materials Korea Ltd., Suwon-Si, Republic of Korea). The solid reagent formic acid was purchased from Merck Ltd., Seoul, Republic of Korea. The QuEChERS extraction kit ($MgSO_4$ 4 g, NaCl 1 g) was purchased from Chiral Technology Korea, Daejeon, Republic of Korea.

2.2. Field Trial

The field trials were carried out from April 2019 to June 2019, and the same cultivar of *Aster scaber* (Asia Seed Co., Ltd., Seoul, Republic of Korea) was used for both fields. Accounting for geographical differences, facility cultivation sites distanced more than 20 km apart (latitudinally) were selected as field trial sites with the locations in Taean-gun, Chungcheongnam province (Field 1), and Gwangyang-si, Jeollanam province (Field 2). Seeding was carried out on 20 June (Field 1) and 28 April (Field 2). The experimental plot was set at 10 m² per repetition and consisted of 3 treatment plots and one non-treatment plot. A small engine knapsack-type sprayer (MSB1015Li, Maruyama, Tokyo, Japan) was utilized for chemical spraying after diluting and preparing the test chemicals in accordance with the safe use standard of agricultural chemicals (Table 1). Samples were taken 0, 1, 2, 3, 5, 7, and 10 days after pesticide application. An appropriate size for the sample was established per day, and an amount of ≥1 kg consisting of at least 12 units was collected. The collected samples were placed in a polyethylene bag, labeled with the chemical name and collection date, stored in an icebox, and immediately delivered to the lab.

Table 1. The formulation of spirodiclofen and tebufenpyrad investigated in the present study.

| Pesticide | Formulation | | | Application | | | | PHI [c] (Days) | MRL [d] (mg/kg) |
	Type	AI [a]	Dilution Rate	Spray No.	Interval (Days)	TSA [b]			
Spirodiclofen	WP [e]	36	4000	2	7	0.9		7	20.0
Tebufenpyrad	EC [f]	10	2000	2	7	0.5		7	1.5

[a] Active ingredient, %; [b] total sprayed amount of pesticides, g; [c] pre-harvest interval; [d] maximum residue limit; [e] wettable powder; [f] emulsifiable concentration.

2.3. Sample Preparation

The samples transported to the laboratory were weighed and then shredded for sample preparation. Before being homogenized with dry ice and a homogenizer, the shredded samples were kept in a refrigerator (below 20 °C) for more than 48 h. Homogenized samples were kept frozen (below 20 °C) until analysis.

2.4. Quality Assurance and Quality Control

The method limit of quantification was determined by instrumental analysis of a standard solution with a signal-to-noise ratio (S/N) of ≥ 10. The LC-MS/MS parameters utilized for the quantitative analysis of spirodiclofen and tebufenpyrad are shown in Table A2. The stock solutions (100 mg/L) of spirodiclofen and tebufenpyrad were prepared in acetonitrile. The prepared stock solution was matrix-matched with *Aster scaber* extract at a ratio of 1:1 to prepare matrix-matched standard solutions of 0.025, 0.05, 0.1, 0.5, 1, and 5.0 mg/L. The calibration curve was drawn by plotting the peak area against the solution concentrations, and the linearity was determined using the regression equation and the coefficient of determination (r^2).

For both pesticides, the recovery rate test was repeated thrice at concentrations of 0.01 and 0.1 mg/kg. An amount of 10 g of the sample was treated with the standard solution to achieve these concentrations, followed by vigorous shaking for 30 min with 10 mL of acetonitrile. The extract was added to the QuEChERS extraction kit ($MgSO_4$ 4 g, NaCl 1 g), shaken, and centrifuged at $4000 \times g$ for 10 min. Then, 1 mL of the supernatant was carefully taken out, mixed with 1 mL of acetonitrile, transferred to a 2 mL autosampler vial, and utilized for the LC-MS/MS analysis (Table A2).

2.5. Storage Stability and Residual Amount per Date

Storage stability was conducted in 3 repetitions by adding 0.1 mg/kg of spirodiclofen and tebufenpyrad standard solutions to untreated samples of 10 g each, uniformly mixing, freezing (below 20 °C), and testing the recovery rates after 162 days and 150 days, respectively. The samples were tested using the same method for the sample analysis that was used in the recovery rate test. The sample analysis used the same preparation and instrumental analysis as the recovery rate test to examine the daily residual amounts of spirodiclofen and tebufenpyrad during the *Aster scaber* cultivation period.

2.6. Calculation of Biological Half-Life and Tolerance Limit of Residue at the Production Stage

The residual amount reduction constant and biological half-lives of spirodiclofen and tebufenpyrad of *Aster scaber* were calculated via regression analysis of their daily residual amounts. After confirming the significance of the regression equation and reduction constant through the *F*-test and the *t*-test, the lower limit of the reduction constant was calculated at the 95% confidence interval. The regression analysis was performed according to the guidelines of the Ministry of Food and Drug Safety, Republic of Korea [13]. The PHRL at the production stage was calculated by estimating the daily residual amount up to 10 days before shipment based on the residual tolerance standards for spirodiclofen and tebufenpyrad of *Aster scaber*.

2.7. Risk Assessment

Risk assessments were conducted on the spirodiclofen and tebufenpyrad residues in *Aster scaber*. The estimated daily intake (EDI) was calculated by multiplying the food consumption and initial pesticide concentration of *Aster scaber* on days 0 and 7, and dividing it by the average body weight [14]. The hazard quotient (HQ) value was calculated using the acceptable daily intake (ADI) and was used for the risk assessment.

3. Results and Discussion

3.1. Temperature, Humidity, and Growth Characteristics within the Facility during Aster scaber Cultivation

During the field-testing period, the mean humidity of Fields 1 and 2 were 76.4 ± 4.7% and 77.4 ± 14.1%, respectively. Similarly, the mean temperatures of Fields 1 and 2 were 16.9 ± 2.6 °C and 17.1 ± 1.8 °C, respectively. The mean weight of 20 plants of *Aster scaber* from day 0 to day 10 after pesticide spraying in Field 1 was 131.0 ± 9.2 g on day 0 and 160.3 ± 4.9 g on day 10. While in Field 2, no significant differences were observed, with 137.3 ± 6.1 g on day 0 and 141.3 ± 5.0 g on day 10. This weight difference was >5 times compared to a previous *Aster scaber* study where the weight of a single unit increased from 0.9 g (day 0) to 4.8 kg (day 10) during the test period (10 days) [15]. In our study, we found that *Aster scaber* cultivated in Fields 1 and 2 only grew to a certain size before slowing down. This is in contrast to cucumbers and broccoli, which grow continuously and rapidly [16].

3.2. Method Validation

The analytical limit of quantification for spirodiclofen and tebufenpyrad in *Aster scaber* was 0.01 mg/kg for both pesticides. The calibration curve of the standard solutions was conducted via regression analysis of the peak area within the concentration range (0.0025–0.5 mg/L), and the linearity within the concentration range was confirmed by the coefficient of determination (r^2). For spirodiclofen, the regression equations calculated through regression analysis were $y = 10{,}601.7x + 30.2$ (Field 1) and $y = 13{,}435.6x + 26.5$ (Field 2), whereas for tebufenpyrad, the equations were $y = 2879.5x - 1.8$ (Field 1) and $y = 2785.7x + 4.0$ (Field 2). The calibration curve equation of spirodiclofen and tebufenpyrad was $y = 10{,}601{,}678.7x + 30{,}166.1$ and $y = 2879{,}505.1x + 1782.6$, respectively. The coefficients of determination of the calibration curves of spirodiclofen and tebufenpyrad were both ≥ 0.999, thereby confirming high linearity. According to the results of the recovery rate tests at concentrations of 0.1 and 1.0 mg/kg, the average recovery rates of spirodiclofen were 100.2% and 104.1%, respectively, whereas those of tebufenpyrad were 89.1% and 103.3%, respectively. The coefficients of variation (CV) were <10%, which was within the acceptable range for the persistence test's analytical method verification criteria (70–110% recovery rate and CV within 20%) for establishing the PHRL (Table 2). During quantitative analysis by LC-MS/MS of spirodiclofen and tebufenpyrad in *Aster scaber*, no interference peaks were observed in the recovery and untreated samples.

Table 2. Recovery (RCV) and storage stability (STR) data for spirodiclofen and tebufenpyrad in *Aster scaber*.

Pesticides	Spiking Levels (mg/kg)		Recovery (%)				CV (%)	MLOQ (mg/kg)
			Replicates			(Mean ± SD)		
			1	2	3			
Spirodiclofen	RCV	0.1	101.4	96.1	103.1	100.2 ± 3.7	3.6	
		1.0	98.0	115.9	98.4	104.1 ± 10.2	9.8	0.01
	STR	1.0	101.8	103.6	100.0	101.8 ± 1.8	1.8	

Table 2. *Cont.*

| Pesticides | Spiking Levels (mg/kg) | | Recovery (%) | | | | CV (%) | MLOQ (mg/kg) |
| | | | Replicates | | | (Mean ± SD) | | |
			1	2	3			
Tebufenpyrad	RCV	0.1	82.0	98.1	87.1	89.1 ± 8.2	9.2	
		1.0	106.0	102.0	101.8	103.3 ± 2.4	2.3	0.01
	STR	1.0	101.4	109.8	101.3	104.2 ± 4.9	4.7	

SD: standard deviation; CV: coefficient of variation; and MLOQ: method limit of quantification.

For the storage stability test conducted at a concentration of 1.0 mg/kg, spirodiclofen showed a recovery rate of 101.8% ± 1.8% after 162 days of storage and tebufenpyrad showed a recovery rate of 104.2% ± 4.9% after 150 days of storage, confirming that there was no degradation or loss of both components during the storage period of the samples.

3.3. Characteristics of Pesticide Residues in Aster scaber

Several studies have reported factors affecting pesticide persistence in crops, including (1) geographical locations and weather conditions of the cultivation area, (2) function, formulation, and application method of pesticides, (3) curvature of the surface of the crop, (4) surface area to weight ratio of the crop, (5) shape and growth rate, (6) amount and shape of villi in crops, (7) composition of the wax layer on the surface of the crop, and (8) the crop cultivation methods [17–20]. In the present study, *Aster scaber* did not show a significant change in weight during the test period; thus, the possibility of the disappearance of pesticides due to the dilution effect (due to weight gain) can be eliminated.

Hong et al. [15] reported that test pesticides have a tendency to disappear during the cultivation period of *Aster scaber* as a result of the dilution effect of the treated chemical caused by the growth rate of *Aster scaber* and its reducing effect on the residual concentration of the pesticide. The dilution effect due to growth was deemed insignificant in this study, as confirmed by the slow growth of the *Aster scaber*.

The difference in residues between Field 1 and Field 2 after treatment with spirodiclofen and tebufenpyrad during the cultivation period of *Aster scaber* was determined. Considering the difference in residue between Field 1 and Field 2, the residue amounts of Field 1 were 1.8 times and 2.9 times higher than that of Field 2 for spirodiclofen and tebufenpyrad, respectively. These results are similar to those of a previous study on *Aster scaber* conducted in the same area where the residue amount of Field 1 was approximately 1.5 times higher than that of Field 2 [16]. This difference in residue amount is considered to have occurred due to the difference in the initial residue amounts of the pesticides. The half-life per region for spirodiclofen was 4.4 and 4.5 days in Fields 1 and 2, respectively. Meanwhile for tebufenpyrad, it was 3.8 and 4.2 days, respectively, indicating no significant difference between the half-lives of the two fields, unlike the residue amounts. It is thought that the half-lives are similar because the crops used in both fields were identical, and the difference between temperature and humidity had little effect on crop growth. In conclusion, even if the initial residue amount differs, if there is no difference in the growth environment, including the crop temperature and humidity, there will be no difference in the pesticide's half-life within the crop's body.

The initial residue amount after spirodiclofen treatment was 14.2 and 8.9 mg/kg in Fields 1 and 2, respectively. Meanwhile for tebufenpyrad, it was 4.9 and 2.0 mg/kg, respectively. This shows that the initial residue amount of spirodiclofen was about 2.9 times higher than that of tebufenpyrad in Field 1 and 4.5 times higher in Field 2. When spraying, the safety standards for both spirodiclofen and tebufenpyrad were the same with up to two treatments, seven days before harvest. Spirodiclofen wettable powder was sprayed at a 4000-fold dilution at 36% content, whereas tebufenpyrad emulsifiable concentration was applied with a 2000-fold dilution at 10% content. Thus, the total amount of spirodiclofen's active ingredient was about 1.8 times greater than that of tebufenpyrad (Table 1). Thus, one of the primary causes of the difference in the initial residue amount of spirodiclofen and tebufenpyrad was determined to be a difference in the total sprayed amount (TSA) of the active ingredient. To account for the difference in TSA, the normalized values (NVs) obtained by dividing the residue amount of each pesticide by TSA were applied (Table 3). The NV of spirodiclofen on day 0 was 15.8 (Field 1) and 9.9 (Field 2), whereas that of tebufenpyrad was 9.8 (Field 1) and 4.0 (Field 2), indicating that the difference in the original residue amount was reduced (concentration not accounting for the amount of active ingredient applied). Although the amount of active ingredient applied was corrected, the difference between the two fields could be due to various factors including the application method, the detailed shape of the crop, and the initial adhesion amount [17–20].

Table 3. Normalized values (NVs) [a] of spirodiclofen and tebufenpyrad in *Aster scaber*.

Pesticides	Fields	Harvest Time (Days after Spraying the Pesticides)							Half-Lives
		0	1	2	3	5	7	10	
Spirodiclofen	1	15.8 [a]	13.6	8.4	4.9	4.0	3.8	3.3	4.4
	2	9.9	6.3	3.7	3.1	2.8	2.4	1.7	4.5
Tebufenpyrad	1	9.8	8.0	6.6	5.6	3.2	2.6	1.6	3.8
	2	4.0	3.2	2.0	1.2	1.2	1.0	0.8	4.2

[a] NVs are calculated as the residue of pesticides (mg/kg)/TSA (total sprayed amount; g).

The regression analysis was utilized to determine the changes in the residue amount over time after treatment with spirodiclofen and tebufenpyrad during the cultivation period of *Aster scaber*. In the present study, during the cultivation period of *Aster scaber*, an exponential decrease in spirodiclofen and tebufenpyrad residue amount was observed. On the 10th day, the last day of harvesting, the initial residue amount of spirodiclofen was 21.1% for Field 1 and 16.9% for Field 2, whereas that of tebufenpyrad was 16.3% for Field 1 and 20.0% for Field 2 (Figure 1). The PHIs of spirodiclofen and tebufenpyrad for *Aster scaber* were seven days. Seven days after the harvest date, the residue amount of spirodiclofen was 3.4 and 2.2 mg/kg for Fields 1 and 2, respectively. In contrast, the residue amount of tebufenpyrad was 1.3 and 0.5 mg/kg for Fields 1 and 2, which were both <20.0 and 1.5 mg/kg (the MRL for each pesticide), respectively.

A regression formula for the daily residual amount in *Aster scaber* calculated through the coefficient of determination (r^2) and simple regression analysis for both pesticides was >0.78, showing a high correlation (Figure 1). The biological half-lives of spirodiclofen in *Aster scaber* were calculated to be 4.4 and 4.5 days in Fields 1 and 2, respectively. In contrast, those of tebufenpyrad were 3.8 and 4.2 days, respectively, indicating no significant difference between the residue loss pattern and biological half-life of both pesticides. These results are consistent with those of a previous study on *Aster scaber* in which the half-lives ranged between 3.6 and 6.7 days [15,16], and are comparable to the half-lives of spirodiclofen in citrus, apple, peaches, and grapes (4.5–11.8 days) [21–24].

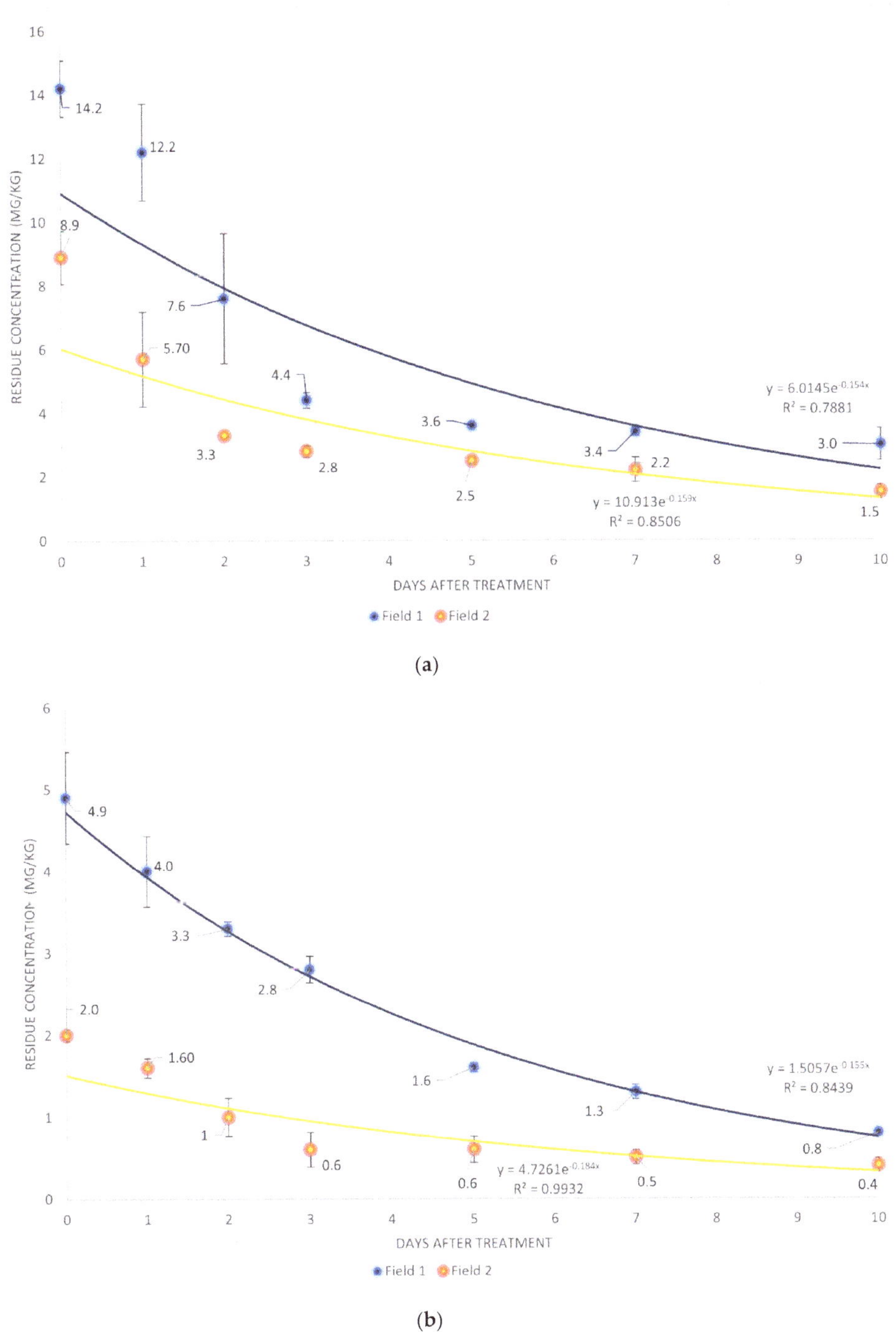

Figure 1. Dissipation kinetics of (**a**) spirodiclofen and (**b**) tebufenpyrad.

3.4. Calculation of the PHRL of Aster scaber

The PHRL refers to the residue amount at a specific point in time prior to harvesting such that the amount of pesticide residue should be below the MRL at the time of harvesting. PHRL is calculated by considering the lower limit of the 95% confidence level of the regression coefficient of the residue amount per day [16]. In the present study, the lower levels of the reduction constants for spirodiclofen were 0.0632 and 0.0723 for Fields 1 and 2, respectively. Whereas for tebufenpyrad, it was 0.1571 and 0.0868 for Fields 1 and 2, respectively. Table 4 displays the computed PHRLs based on these values. The PHRLs (10 days prior to harvest) of spirodiclofen for *Aster scaber* were 37.6 and 41.2 mg/kg for Fields 1 and 2, respectively. Meanwhile for tebufenpyrad, it was 7.2 and 3.6 mg/kg for Fields 1 and 2, respectively.

Table 4. Recommended pre-harvest residue limits (PHRLs) of spirodiclofen and tebufenpyrad in *Aster scaber*.

Pesticide		Recommended PHRLs (mg/kg)				MRLs (mg/kg)
		10 Days Before Harvesting	7 Days Before Harvesting	5 Days Before Harvesting	3 Days Before Harvesting	
Spirodiclofen	Field 1	37.6	31.1	27.4	24.2	20.0
	Field 2	41.2	33.2	28.7	24.8	
Tebufenpyrad	Field 1	7.2	4.5	3.3	2.4	1.5
	Field 2	3.6	2.7	2.3	2.0	

3.5. Risk Assessment

A risk assessment was conducted on the residues of spirodiclofen and tebufenpyrad in *Aster scaber*. Table 5 shows the results of the risk assessments of spirodiclofen and tebufenpyrad on days 0 and 7 after spraying according to the ingestion amount of *Aster scaber*. If the HQ is more than 100%, it indicates a high risk. In the case of spirodiclofen, the HQs on day 0 of the application were 178.9% (Field 1) and 112.1% (Field 2), whereas the HQs on day 7 of the application were 42.8% (Field 1) and 27.7% (Field 2). For tebufenpyrad, the HQs on day 0 of the application were 61.7% (Field 1) and 25.2% (Field 2), whereas the HQs on day 7 of the application were 16.4% (Field 1) and 6.3% (Field 2). On day 0, the hazard quotient (HQ) for spirodiclofen in both fields exceeded 100%, indicating a risk; however, on day 7 of the PHI period, the HQ decreased below 50%, indicating that it is not a high risk. In the case of tebufenpyrad, the HQ on days 0 and 7 of the application was less than 70%, making it less risky than spirodiclofen. However, the above HQ results were higher than the HQ range of 0.1%–3.9% found in a previous study on *Aster scaber* [25]. The sample collection period in their study was one year, obtained from large retailers and wholesale markets for agriculture products. Therefore, in agreement with previous studies, the present study suggests that consumption of *Aster scaber* in the production stage (0 and 7 days) has a lower risk.

Table 5. Risk assessment of spirodiclofen and tebufenpyrad in *Aster scaber*.

	Pesticide		Residue Value (mg/kg)	ADI [a] (mg/kg bw/Day)	EDI [b] (mg/kg bw/Day)	HQ [c] (%)
Spirodiclofen	Day 0	Field 1	14.2	0.01	0.18	178.9
		Field 2	8.9	0.01	0.11	112.1
	Day 7 day	Field 1	3.4	0.01	0.004	42.8
		Field 2	2.2	0.01	0.003	27.7
Tebufenpyrad	Day 0	Field 1	4.9	0.01	0.006	61.7
		Field 2	2.0	0.01	0.003	25.2
	Day 7	Field 1	1.3	0.01	0.002	16.4
		Field 2	0.5	0.01	0.001	6.3

[a] Acceptable daily intake; [b] estimated daily intake; [c] hazard quotient; HQ = EDI/ADI.

4. Conclusions

The decreasing trend and residual characteristics of spirodiclofen and tebufenpyrad pesticides were identified in the *Aster scaber* production stage, and a risk assessment was conducted. While the initial residue amount differed in the residue reduction trend of the two fields, there was no difference in the half-lives because of similar cultivation conditions. All HQs were <100% in the risk assessment conducted using the residual amount on day 7 after pesticide application, corresponding to the PHI period of both pesticides, indicating that the risk is considered low. However, in the case of spirodiclofen, the HQs on days 0 and 7 after application were >100% and >25%, respectively, which could be considered a risk.

Author Contributions: Conceptualization, R.K.S., R.K., K.L. and J.K.; methodology, R.K. and J.K.; software, R.K.S.; validation, R.K., K.L. and J.K.; formal analysis, R.K. and D.A.; investigation, R.K. and D.A.; resources, H.-R.C. and J.-H.L.; data curation, R.K. and D.A.; writing—original draft preparation, H.-R.C., J.-H.L. and Y.S.; writing—review and editing, R.K.S. and J.-H.L.; visualization, J.-H.L.; supervision, H.-R.C. and J.-H.L.; project administration, H.-R.C. and J.-H.L.; funding acquisition, H.-R.C. and J.-H.L. All authors have read and agreed to the published version of the manuscript.

Funding: This research was supported by the Ministry of Food and Drug Safety, Republic of Korea (grant number: 17162MFDS010).

Data Availability Statement: Data will be made available on request.

Acknowledgments: This paper was supported by the KU research professor program of Konkuk University, Seoul, Republic of Korea.

Conflicts of Interest: The funders had no role in the design of the study; in the collection, analyses, or interpretation of the data; in the writing of the manuscript; or in the decision to publish the results.

Appendix A

Table A1. Chemical Properties of Spirodiclofen and Tebufenpyrad [7].

Pesticide	Spirodiclofen	Tebufenpyrad
Chemical structure		
Vapor pressure	$<3.0 \times 10^{-4}$ mPa (20 °C)	$<1.0 \times 10^{-2}$ mPa (25 °C)
log K_{ow}	5.1	4.9
Water solubility	0.19 mg/L (20 °C)	2.61 mg/L (25 °C)
Stability (DT$_{50}$)	52.1 day (hydrolysis)	Stable to hydrolysis

Table A2. Analytical conditions of spirodiclofen and tebufenpyrad in *Aster scaber*.

Pesticides	Spirodiclofen			Tebufenpyrad		
Instrument	Shimadzu LC-MS 8045 (Shimadzu, Tokyo, Japan)					
Column	Kinetex C18 (100 × 2.1 mm, 2.6 μm, Phenomenex)					
Mobile phase	A: 0.1% formic acid in distilled water B: 0.1% formic acid in acetonitrile					
	Time (min)	A (%)	B (%)	Time (min)	A (%)	B (%)
	0.5	60	40	3	20	80
	2	5	95	3.5	10	90
	5	5	95	5.5	10	90
	6	60	40	6	20	80
	7	60	40	7	20	80
Flow rate	0.2 mL/min					
Injection volume	2 μL					
Retention time	4.19 min			1.47 min		
Detector	Triple-quadruple system					
Multiple reaction monitoring (MRM) parameters	Precursor ion > Product ion (collision energy voltage) Quantifier ion; 410.9 > 71.2 (−23.0) Qualifier ion; 410.9 > 313.0 (−13.0)			Precursor ion > Product ion (collision energy voltage) Quantifier ion; 334.0 > 145.1 (−29.0) Qualifier ion; 334.0 > 117.1 (−35.0)		

References

1. Fenoll, J.; Hellin, P.; Camacho, M.D.; Lopez, J.; Gonzalez, A.; Lacasa, A.; Flores, P. Dissipation rates of procymidone and azoxystrobin in greenhouse grown lettuce and under cold storage conditions. *Int. J. Environ. Anal. Chem.* **2008**, *88*, 737–746. [CrossRef]
2. Kim, J.-E.; Choi, T.-H. Behavior of Synthetic Pyrethroid Insecticide Bifenthrin in Soil Environment I) Degradation Pattern of Bifenthrin and Cyhalothrin in Soils and Aqueous Media. *Korean J. Environ. Agric.* **1992**, *11*, 116–124.
3. Krol, W.J.; Arsenault, T.L.; Pylypiw, H.M., Jr.; Incorvia Mattina, M.J. Reduction of pesticide residues on produce by rinsing. *J. Agric. Food Chem.* **2000**, *48*, 4666–4670. [CrossRef] [PubMed]
4. Abdel Ghani, S.B.; Abdallah, O.I. Method validation and dissipation dynamics of chlorfenapyr in squash and okra. *Food Chem.* **2016**, *194*, 516–521. [CrossRef] [PubMed]
5. Torres, C.M.; Pico, Y.; Manes, J. Determination of pesticide residues in fruit and vegetables. *J. Chromatogr. A* **1996**, *754*, 301–331. [CrossRef]
6. Lee, D.; Jo, H.; Jeong, D.; Goo, Y.; Hwang, M.; Kang, N.; Kang, K.; Kim, J. Residual characteristics of pesticides used for powdery mildew control on greenhouse strawberry. *J. Agric. Life Sci.* **2018**, *52*, 99–106. [CrossRef]
7. Ouyang, Y.; Montez, G.H.; Liu, L.; Grafton-Cardwell, E.E. Spirodiclofen and spirotetramat bioassays for monitoring resistance in citrus red mite, *Panonychus citri* (Acari: Tetranychidae). *Pest Manag. Sci.* **2012**, *68*, 781–787. [CrossRef]
8. Rauch, N.; Nauen, R. Spirodiclofen resistance risk assessment in *Tetranychus urticae* (Acari: Tetranychidae): A biochemical approach. *Pestic. Biochem. Physiol.* **2002**, *74*, 91–101. [CrossRef]
9. Korea Crop Protection Association (KCPA). Crop Protection Guidelines. Korea Crop Protection Association, Seoul. 2020. Available online: https://www.koreacpa.org/ko/use-book/search/ (accessed on 3 March 2022).
10. Kim, E.H.; Shim, Y.Y.; Lee, H.I.; Lee, S.; Reaney, M.J.T.; Chung, M.J. Astragalin and Isoquercitrin Isolated from *Aster scaber* Suppress LPS-Induced Neuroinflammatory Responses in Microglia and Mice. *Foods* **2022**, *11*, 1505. [CrossRef]
11. Kim, G.-M.; Hong, J.-Y.; Woo, S.-Y.; Lee, H.-S.; Choi, Y.-J.; Shin, S.-R. Antioxidant activities of ethanol extracts of *Aster scaber* grown in wild and culture field. *Korean J. Food Preserv.* **2015**, *22*, 567–576. [CrossRef]
12. Tomlin, C.D. *The Pesticide Manual: A World Compendium*; British Crop Production Council: Farnham, UK, 2009.
13. Chang, H.-R.; You, J.-S.; Ban, S.-W. Residue Dissipation Kinetics and Safety Evaluation of Insecticides on Strawberry for the Harvest Periods in Plastic-covered Greenhouse Conditions. *Korean J. Environ. Agric.* **2020**, *39*, 122–129. [CrossRef]
14. Park, B.K.; Kwon, S.H.; Yeom, M.S.; Joo, K.S.; Heo, M.J. Detection of pesticide residues and risk assessment from the local fruits and vegetables in Incheon, Korea. *Sci. Rep.* **2022**, *12*, 9613. [CrossRef] [PubMed]
15. Hong, J.-H.; Lim, J.-S.; Lee, C.-R.; Han, K.-T.; Lee, Y.-R.; Lee, K.-S. Study of pesticide residue allowed standard of methoxyfenozide and novaluron on *Aster scaber* during cultivation stage. *Korean J. Pestic. Sci.* **2011**, *15*, 8–14.
16. Lee, S.; Ko, R.; Lee, K.; Kim, J.; Kang, S.; Lee, J. Dissipation patterns of acrinathrin and metaflumizone in Aster scaber. *Appl. Biol. Chem.* **2022**, *65*, 14. [CrossRef]
17. Ghadiri, H.; Rose, C.W.; Connell, D.W. Degradation of organochlorine pesticides in soils under controlled environment and outdoor conditions. *J. Environ. Manag.* **1995**, *43*, 141–151. [CrossRef]
18. Kim, J.; Song, B.; Chun, J.; Im, G.; Im, Y. Effect of sprayable formulations on pesticide adhesion and persistence in several crops. *Korean J. Pestic. Sci.* **1997**, *1*, 35–40.

19. Lee, H.-S.; Park, Y.-W. Antioxidant activity and antibacterial activities from different parts of broccoli extracts under high temperature. *J. Korean Soc. Food Sci. Nutr.* **2005**, *34*, 759–764.
20. Poulsen, M.E.; Wenneker, M.; Withagen, J.; Christensen, H.B. Pesticide residues in individual versus composite samples of apples after fine or coarse spray quality application. *Crop Prot.* **2012**, *35*, 5–14. [CrossRef]
21. Sun, D.; Zhu, Y.; Pang, J.; Zhou, Z.; Jiao, B. Residue level, persistence and safety of spirodiclofen-pyridaben mixture in citrus fruits. *Food Chem.* **2016**, *194*, 805–810. [CrossRef]
22. Sun, H.; Liu, C.; Wang, S.; Liu, Y.; Liu, M. Dissipation, residues, and risk assessment of spirodiclofen in citrus. *Environ. Monit. Assess.* **2013**, *185*, 10473–10477. [CrossRef]
23. Bai, Y.; Xu, P.; Gao, X.; Sun, W.; Zhang, H. Determination of residual spirodiclofen in apple and soil. *J. Anal. Sci.* **2009**, *25*, 229–231.
24. Peng-Jun, X.; Xiao-Sha, G.; Bu, T.; Hong-Yan, Z.; Shu-Ren, J. Determianion of Spirodiclofen Residue in Nine Fruits by Gas Chromatography-Quadrupole Mass Spectrometry with Dispersive Solid Phase Extraction. *Chin. J. Anal. Chem.* **2008**, *36*, 1515–1520.
25. Park, B.K.; Jung, S.H.; Kwon, S.H.; Kim, S.H.; Yeo, E.Y.; Yeom, M.S.; Seo, S.J.; Joo, K.S.; Heo, M.J.; Hong, G.P. Health risk associated with pesticide residues in vegetables from Incheon region of Korea. *Environ. Sci. Pollut. Res.* **2022**, *29*, 65860–65872. [CrossRef] [PubMed]

Article

Magnetic Composite Based on Carbon Nanotubes and Deep Eutectic Solvents: Preparation and Its Application for the Determination of Pyrethroids in Tea Drinks

Xiaodong Huang, Huifang Liu, Xiaomin Xu, Ge Chen, Lingyun Li, Yanguo Zhang, Guangyang Liu and Donghui Xu *

Institute of Vegetables and Flowers, Chinese Academy of Agricultural Sciences, Key Laboratory of Vegetables Quality and Safety Control, Laboratory of Quality & Safety Risk Assessment for Vegetable Products, Ministry of Agriculture and Rural Affairs of China, Beijing 100081, China
* Correspondence: xudonghui@caas.cn; Tel.: +86-10-82106963

Abstract: In this study, a novel composite material prepared by using deep eutectic solvent (tetrabutylammonium chloride-dodecanol, DES_5) functionalized magnetic $MWCNTs-ZIF-8$ ($MM/ZIF-8@DES_5$) was employed as an adsorbent for the magnetic solid-phase extraction of six pyrethroids from tea drinks. The characterization results show that $MM/ZIF-8@DES_5$ possessed sufficient specific surface area and superparamagnetism, which could facilitate the rapid enrichment of pyrethroids from tea drink samples. The results of the optimization experiment indicated that DES_5, which comprised tetrabutylammonium chloride and 1-dodecanol, was selected for the next experiment and that the adsorption properties of $MM/ZIF-8@DES_5$ were higher than those of $MM/ZIF-8$ and M-MWCNTs. The validation results show that the method has a wide linear range (0.5–400 $\mu g\ L^{-1}$, $R^2 \geq 0.9905$), low LOD (0.08–0.33 $\mu g\ L^{-1}$), and good precision (intra-day RSD $\leq 5.6\%$, inter-day RSD $\leq 8.6\%$). The method was successfully applied to the determination of pyrethroids in three tea drink samples.

Keywords: magnetic solid-phase extraction; deep eutectic solvent; $ZIF-8$; pyrethroids; tea drinks

Citation: Huang, X.; Liu, H.; Xu, X.; Chen, G.; Li, L.; Zhang, Y.; Liu, G.; Xu, D. Magnetic Composite Based on Carbon Nanotubes and Deep Eutectic Solvents: Preparation and Its Application for the Determination of Pyrethroids in Tea Drinks. *Foods* **2023**, *12*, 8. https://doi.org/10.3390/foods12010008

Academic Editor: Amin Mousavi Khaneghah

Received: 18 November 2022
Revised: 9 December 2022
Accepted: 19 December 2022
Published: 20 December 2022

1. Introduction

Magnetic solid-phase extraction (MSPE), evolved from traditional solid-phase extraction (SPE), has attracted extensive attention as a preconcentration technique because of its ease of use, low consumption of organic solvents, and time- and cost-efficiency [1]. In a typical MSPE procedure, a small quantity of magnetic sorbent material is directly exposed to a sample solution followed by extraction processing until an adsorption equilibrium is achieved. Then, the sorbents containing the target analytes are retrieved under an external magnetic field; consequently, MSPE exhibits superior extraction efficiency compared with traditional SPE [2]. The sorbent always plays a key role in the MSPE technique, and numerous materials have been prepared as magnetic adsorbents for MSPE, such as carbon nanomaterials, polymers, molecularly-imprinted materials, and porous materials [3–6].

Multiwalled carbon nanotubes (MWCNTs), a class of tubular carbon nanomaterials based on layers of seamlessly rolled up graphene sheets, have typically been integrated with Fe_3O_4 nanoparticles and used as magnetic adsorbents for MSPE. MWCNTs exhibit excellent extraction performance for different analytes due to their remarkable chemical properties, stability, surface area and mechanical strength [7]. In recent years, a new trend has emerged that expands the sample preparation applications of Fe_3O_4-MWCNTs (M-MWCNTs) by integrating them with metal–organic frameworks (MOFs) to form functional composites [8].

Zeolitic imidazolate frameworks (ZIFs), as a new family of MOFs, have crystalline three-dimensional frameworks in which transition metal ions (especially as Zn^{2+} Cu^{2+} and Co^{2+}) are linked by imidazolate-type organic linkers [9]. ZIFs have found several applications in the adsorption of pollutants from aqueous samples due to their special properties of aqueous and thermal stability, and adsorption capacity [10]. ZIF$-$8, composed of Zn^{2+} and 2-methylimidazole, has received wide attention in sample preparation, especially for ZIF$-$8 composites prepared by integration with other functional materials [11]. To promote the extraction performance of ZIF$-$8 composites for pesticide residues in aqueous samples, deep eutectic solvents (DESs) can be used as an effective reagent to modify ZIF$-$8 composites [12]. DESs, traditionally prepared by using a hydrogen bond donor (HBD) and hydrogen bond acceptor (HBA) under the action of an intermolecular hydrogen bond, have several fascinating features including low melting point (<100 $^{\circ}$C), excellent heat stability, extremely low vapor pressure, naturally degradable, easy to prepare, low cost, and tunable miscibility in water [13,14]. As green and effective reagents, DESs can be selected as an effective extractant in the extraction of active ingredients from plants, heavy metals, environmental pollutants, and pesticides [15,16].

Pyrethroids, as a class of efficient bioderived broad-spectrum insecticides including 42 substances, are the fourth group of insecticides on the basis of the WHO classification [17]. Pyrethroid insecticides have been widely used in agriculture because of their high efficacy, low acute oral toxicity, and harmony with the environment [18,19]. However, long-term and extensive application of pyrethroids could threaten the environment and food chains [20]. Moreover, pyrethroids possess characteristics of bioaccumulation in marine mammals and humans, and their acceptable daily intake values range from 0.02–0.07 mg kg^{-1} day^{-1} and no observed adverse effect level values range from 1–7 mg kg^{-1} day^{-1} [21]. Nowadays, countries in the world have established maximum residue limits (MRLs) of pyrethroids in tea products, for instance MRLs set at 0.1–50 mg kg^{-1} in China (GB 2763 National food safety standard) [22]. Therefore, it is necessary to establish a sensitive and reliable determination method for monitoring pyrethroid pesticides.

This study aimed to prepare a magnetic composite based on carbon nanotubes and deep eutectic solvents and employ this as an adsorbent for the MSPE of pyrethroids from tea (*Camellia sinensis* L.) drink samples. The modification of M-MWCNTs/ZIF$-$8 (MM/ZIF$-$8) by the use of ionic liquids and the functionalization of MOFs by the introduction of DESs had been reported. However, there are no reports of tetrabutylammonium chloride-dodecanol-based DES$_5$ being used to functionalize a MM/ZIF$-$8 composite or their use as an adsorbent for the MSPE of pyrethroids. The MM/ZIF$-$8@DES was obtained and characterized, and the pretreatment technique parameters were optimized. Finally, the established MM/ZIF$-$8@DES-based sample pretreatment technique was successfully used to extract and determine the amounts of six pyrethroids in tea drinks.

2. Materials and Methods

2.1. Preparation of Magnetic Materials

2.1.1. Preparation of MM/ZIF$-$8

M-MWCNTs were prepared following a modified chemical co-precipitation method reported previously [23]. First, 0.2 g of MWCNTs powder (95%, inside diameter of 3–5 nm, length of 50 μm, Aladdin Co., Shanghai, China) was suspended in ultrapure water (240 mL) in a three-necked flask under ultrasonic irradiation for 1 h. After that, 1.8 g of FeCl$_3$·6H$_2$O (Aladdin Co., Shanghai, China) and 0.8 g of FeCl$_2$·4H$_2$O (Aladdin Co., Shanghai, China) were added to the flask with mechanical stirring for 30 min at 80 $^{\circ}$C. Then, 10 mL of NH$_3$·H$_2$O (28%, Aladdin Co., Shanghai, China) was slowly transferred into the flask, followed by another 30 min of incubating. Finally, the M-MWCNT material was gathered by magnetic adsorption and cleaned for 3 rounds to remove unreacted chemicals with ultrapure water and anhydrous ethanol (analytical grade, Beijing Bailingwei Science and Technology Co., Beijing, China), sequentially.

The preparation of MM/ZIF$-$8 was conducted according to our previously published method [24]. First, all of the obtained M-MWCNTs from the former step were suspended in anhydrous ethanol (140 mL), which contained 230 μL of mercaptoacetic acid (analytical grade, Beijing Bailingwei Science and Technology Co., Beijing, China), followed by mechanical stirring for 60 min at room temperature. Then, the reacted mixture was separated with the help of an external magnet and cleaned for 3 rounds to remove unreacted chemicals with ultrapure water and anhydrous ethanol, sequentially. After that, the synthetic product was added into the mixture of anhydrous ethanol-ultrapure water (1:1, v/v, 240 mL), which also contained 0.26 g of $ZnSO_4·7H_2O$ (Aladdin Co., Shanghai, China), and the mixture was stirred for 1.5 h. Then, 0.84 g of 2-methylimidazole (Aladdin Co., Shanghai, China) was dissolved in anhydrous ethanol (20 mL) and transferred into the above mixture, followed by another 8 h of mechanical stirring. Finally, the obtained synthetic products were acquired by means of magnetic adsorption and cleaned for several rounds to remove unreacted chemicals using anhydrous ethanol and ultrapure water, sequentially. The synthetic MM/ZIF−8 was dried in a vacuum drying oven at 60 °C for a whole day.

2.1.2. Preparation of MM/ZIF−8@DES

The preparation of six kinds of DESs were carried out as follows: HBAs and HBDs were blended in a 1:2 molar ratio, followed by constant mechanical stirring at 40 °C until a homogeneous solution was formed. The prepared DESs products were based on two kinds of quaternary ammonium salts and three different HBDs according to a specific molar ratio (Table 1).

Table 1. List of synthetic components of different DESs.

DES	HBA	HBD	Molar Ratio of HBA to HBD
DES$_1$	1-methyl-3-octyl imidazolium chloride [a]	1-undecanol [b]	1:2
DES$_2$	1-methyl-3-octyl imidazolium chloride	1-dodecanol [c]	1:2
DES$_3$	1-methyl-3-octyl imidazolium chloride	1-tridecanol [d]	1:2
DES$_4$	tetramethylammonium chloride [e]	1-undecanol	1:2
DES$_5$	tetramethylammonium chloride	1-dodecanol	1:2
DES$_6$	tetramethylammonium chloride	1-tridecanol	1:2

[a,b,c,d,e] Five kinds of reagents were analytical grade; obtained from Beijing Bailingwei Science and Technology Co. (Beijing, China).

MM/ZIF−8@DES was fabricated in accordance with a published method with a slight modification [25]. First, Different amounts of MM/ZIF−8 (0.25, 0.4, 0.5, 0.75, 1.0, and 1.25 g) were added to the 4 mL of methanol (HPLC grade, Sigma-Aldrich, STL, USA) solution, which including 0.5 g of DES. Then, the mixtures were incubated for 1 h under ultrasonic conditions at ambient temperature. After that, the synthetic MM/ZIF−8@DES was obtained via magnetic adsorption, followed by 3 rounds of anhydrous ethanol washes, and vacuum drying at 60 °C for a whole day.

2.2. The MSPE Procedure

A total of 100 mg L^{-1} of stock standard solution was prepared as follows: First, 1 mL of pyrethroid standard solutions (1000 mg L^{-1}, Agro-Environmental Protection Institute, Tianjin, China) of cyhalothrin, cyfluthrin, cypermethrin, flucythrinate, fenvalerate and fluvalinate, were added into a 10 mL volumetric flask, respectively. After that, the mixture was diluted with HPLC grade methanol (Sigma-Aldrich, STL, USA) to volume followed by hand shaking for 20 s. Then, the prepared stock solution was transferred into a brown color screw sample bottle and stored at −20 °C.

The MSPE was programed as follows (Figure 1): First, 6 mg of MM/ZIF−8@DES$_5$ was suspended into 5 mL of sample solution, which contained a certain concentration of pyrethroids, followed by 10 min of shaking to reach an adsorption equilibrium state. Then, the mixed solution was placed in an outside magnetic field to remove the supernatant. After that, 3 min of vortex desorption was carried out by adding 3 mL of ethyl acetate (HPLC grade, Sigma-Aldrich, STL, USA) to the adsorbent. The supernatant desorption solution was collected by an outside magnetic field, followed by an evaporation treatment until dry via gentle nitrogen-blowing at 40 °C. Finally, the residue component was dissolved once again with 0.5 mL of acetone (HPLC grade, Sigma-Aldrich, STL, USA), a 1.0 µL portion of which was analyzed by gas chromatography-tandem mass spectrometry (GC-MS/MS, model GCMS-TQ 8040, Shimadzu, Kyoto, Japan). The operational parameters for GC-MS/MS and the detailed multiple reaction monitoring (MRM) transitions for the six pyrethroids are presented in Tables 2 and 3, respectively.

Table 2. Operational parameters for GC-MS/MS.

	GC Specification		**MS Specification**	
Column	Rtx-5Ms capillary column (0.25 mm (id) × 30 m, 0.25 µm film thickness, Restek, Bellefonte, PA, USA)		Interface temperature	300 °C
Column oven temp	40 °C (4 min), 40–125 °C at 25 °C min^{-1}, 125–300 °C at 10 °C min^{-1}, and finally held for 6 min. The total run time was 21 min		Ion source temperature	200 °C
Carrier gas and column flow	Helium flow rate = 1.0 mL min^{-1}		Measurement mode	MRM
Injection	1.0 µL, splitless mode		—	—

Table 3. Acquisition and chromatographic parameters of the six pyrethroids.

Analytes	t_R (min)	MRM1 (*m/z*)	CE1 (eV)	MRM2 (*m/z*)	CE2 (eV)
Cyhalothrin-1	18.785	197.0 > 161.0	8	197.0 > 141.0	12
Cyhalothrin-2	18.962	197.0 > 161.0	8	197.0 > 141.0	12
Cyfluthrin-1	20.304	226.1 > 206.1	14	226.1 > 199.1	6
Cyfluthrin-2	20.398	226.1 > 206.1	14	226.1 > 199.1	6
Cyfluthrin-3	20.461	226.1 > 206.1	14	226.1 > 199.1	6
Cyfluthrin-4	20.501	226.1 > 206.1	14	226.1 > 199.1	6
Cypermethrin-1	20.630	163.1 > 127.1	6	163.1 > 91.0	14
Cypermethrin-2	20.733	163.1 > 127.1	6	163.1 > 91.0	14
Cypermethrin-3	20.793	163.1 > 127.1	6	163.1 > 91.0	14
Cypermethrin-4	20.831	163.1 > 127.1	6	163.1 > 91.0	14
Flucythrinate-1	20.794	199.1 > 157.1	10	199.1 > 107.1	22
Flucythrinate-2	20.985	199.1 > 157.1	10	199.1 > 107.1	22
Fenvalerate-1	21.430	419.1 > 225.1	6	419.1 > 167.1	12
Fenvalerate-2	21.640	419.1 > 225.1	6	419.1 > 167.1	12
Fluvalinate-1	21.540	250.1 > 55.0	20	250.1 > 200.0	20
Fluvalinate-2	21.600	250.1 > 55.0	20	250.1 > 200.0	20

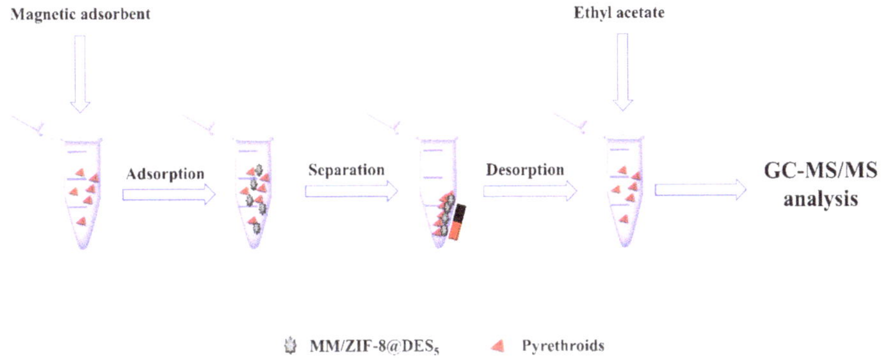

Figure 1. Schematic procedure for the proposed MSPE method.

3. Results and Discussions

3.1. Characterization of Magnetic Materials

The morphologies of M-MWCNTs and MM/ZIF−8@DES$_5$ were studied by transmission electron microscopy (TEM, performed with JEM-200CX transmission electron microscope, JEOL, Tokyo, Japan) and scanning electron microscopy (SEM, performed with JSM-6300 scanning electron microscope, JEOL, Tokyo, Japan), respectively. As shown in Figure 2A, MWCNTs present a typical tubular shape, and Fe$_3$O$_4$ nanoparticles were adhered to the surface of MWCNTs with slight agglomeration. As shown in Figure 2B, MM/ZIF−8@DES$_5$ reveals a highly-porous block-shaped structure with a rough surface, which indicates a good adsorption prospect for the enrichment of pesticides. The crystal structures of the obtained materials were identified by X-ray diffractometer (XRD, performed with D8 Advance X-ray powder diffractometer, Bruker, Karlsruhe, Germany), and the results are shown in Figure 2C. Several primary diffraction peaks for Fe$_3$O$_4$, appearing at 21.3°, 35.2°, 41.5°, 63.2°, 67.4°, and 74.5°, can be seen clearly in three M-MWCNTs-based materials. These results indicated that Fe$_3$O$_4$ nanoparticles were retained during the formation of the composite materials. Peaks of 10–25° and 28–34° demonstrated that the MM/ZIF−8@DES$_5$ was successfully prepared [26].

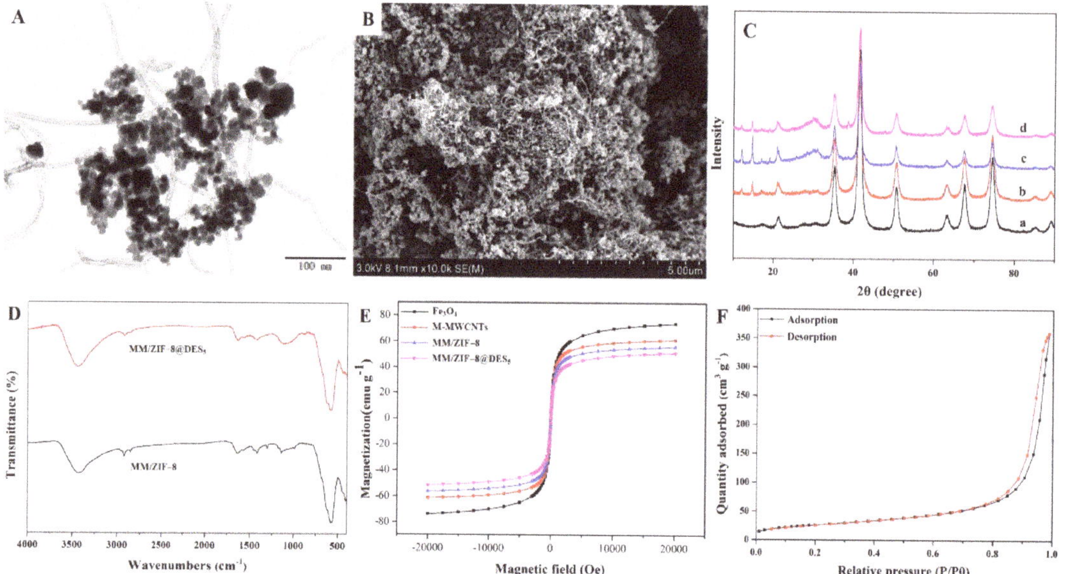

Figure 2. Characterization of the prepared materials: (**A**) TEM image of M-MWCNTs; (**B**) SEM image of MM/ZIF−8@DES$_5$; (**C**) XRD patterns of (a) Fe$_3$O$_4$, (b) M-MWCNTs, (c) MM/ZIF−8, and (d) MM/ZIF−8@DES$_5$; (**D**) FT-IR spectra of magnetic composites; (**E**) Magnetic curves of obtained materials; (**F**) N$_2$ adsorption-desorption isotherms (obtained at 300 K) of MM/ZIF−8@DES$_5$.

Figure 2D exhibits the Fourier transform infrared (FT-IR, performed with FT-IR-8400 spectrometer, Shimadzu, Kyoto, Japan) spectra of the obtained products. The adsorption bands at 572 cm^{-1} were originated from Fe-O stretching vibration, indicating that the successful encapsulation of Fe$_3$O$_4$ into magnetic materials. Meanwhile, the adsorption peak at 1534 cm^{-1} corresponds to the cylinder-like carbon structure of MWCNTs, and the characteristic bands at 432, 811–1360, and 1418 cm^{-1} probably because of the Zn-N stretching vibration and at 911 cm^{-1} for the C-N stretching vibration from the imidazole ring. These results suggest the successful preparation of MM/ZIF−8 [23]. The peak at 1475 cm^{-1} can be attributed to the C-N stretching vibration of DES [27]. Furthermore, the characteristic absorption peaks of ZIF−8 in the spectra of MM/ZIF−8@DES$_5$ were weaker than those in MM/ZIF$-$8, which affected the FT-IR spectral scanning of ZIF$-$8 because of the coating of DES. The above results indicate that MM/ZIF$-$8@DES$_5$ was successfully synthesized.

Magnetic characteristics of the prepared magnetic composites were confirmed by vibrating sample magnetometry (VSM, performed with 7410 magnetometer, Lake Shore, Columbus, USA) (Figure 2E). The magnetic hysteresis loops of Fe$_3$O$_4$, M-MWCNTs, MM/ZIF$-$8, and MM/ZIF$-$8@DES$_5$ all presented an S-like curve, and the saturation magnetizations of them were 75.5, 61.3, 56.1, and 51.3 emu g^{-1}, respectively. Moreover, values of coercivity and remanence for the above materials are negligible. These results show that all prepared materials are superparamagnetic and are capable of rapidly separating with an external magnet [12].

The porosity of MM/ZIF$-$8@DES$_5$ was investigated by an N$_2$ adsorption-desorption isotherm (performed with ASAP 2020 surface area and porosity analyzer at 300 K, Micromeritics, Norcross, USA). As presented in Figure 2F, the comprehensive pattern of curves reveals that the N$_2$ adsorption increased slightly at low relative pressures and sharply increased at high relative pressure. These results suggest that the pore size of MM/ZIF$-$8@DES$_5$ is in the range of the mesoporous to microporous scale [24]. Furthermore, the Brunauer–Emmett–Teller surface area and the pore volume were 133.68 m^2 g^{-1} and 0.574 mL g^{-1}, respectively. All the above results illustrate that MM/ZIF$-$8@DES$_5$ possesses an acceptable surface area and total volume, which are conducive to the adsorption of pyrethroids.

3.2. The Optimization of the MSPE Parameters

To achieve a satisfactory extraction efficiency for the developed MSPE technique, a single-factor experimental design was adopted to optimize the type and quantity of DES, and the adsorption and desorption conditions.

3.2.1. Selection of Type and Quantity of DES$_5$

To examine the effect of different DESs on the extraction performance, six types of DES with different HBDs and HBAs were prepared and used to coat the surface of the magnetic composites. As shown in Figure 3A, the extraction effects of DES$_4$, DES$_5$, and DES$_6$ were more superior to those of DES$_1$, DES$_2$, and DES$_3$. In DES$_1$, DES$_2$, and DES$_3$, the extraction performance for pyrethroids promoted with the alkyl chain length of the HBD increased, which was ascribed to the principle of similar phase-dissolving. In DES$_4$, DES$_5$, and DES$_6$, the extraction performance was not related to the length of the alkyl chain of the HBD. With consideration of the smallest relative standard deviations (RSD), the optimal modifier chosen was DES$_5$.

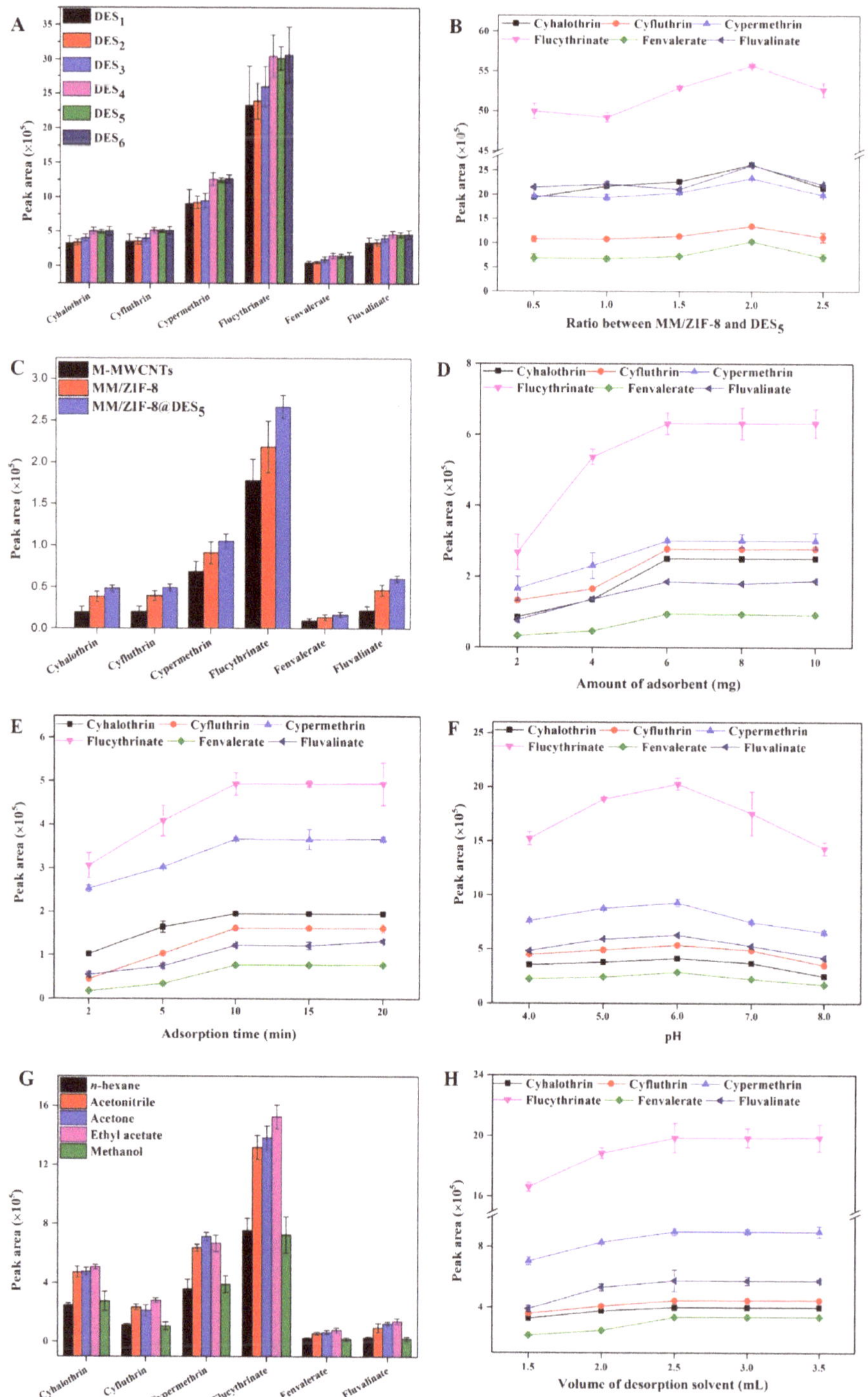

Figure 3. Effect of conditions on the MSPE of pyrethroids: (**A**) type of DES; (**B**) mass ratio of MM/ZIF−8, and DES$_5$; (**C**) type of adsorbent; (**D**) amount of adsorbent; (**E**) adsorption time; (**F**) pH value of sample solution; (**G**) type of desorption solvent; (**H**) desorption time.

The mass ratio of MM/ZIF$-$8 to DES has a significant influence on the adsorption of the analytes. To achieve a positive extraction efficiency, several mass ratios (0.5:1, 1:1, 1.5:1, 2:1, and 2.5:1) of MM/ZIF$-$8 to DES_5 were investigated (Figure 3B). The extraction performance for analytes were optimal when the mass ratio was 2:1. Therefore, this ratio was selected as the optimal constituent of MM/ZIF$-$8@DES_5.

3.2.2. Effect of Type and Dosage of Adsorbent

To identify the extraction performance of prepared magnetic materials for pyrethroids, three composites, including MM/ZIF$-$8@DES_5, MM/ZIF$-$8, and M-MWCNTs were selected as potential magnetic adsorbents for the MSPE procedure (Figure 3C). Clearly, MM/ZIF$-$8@DES_5 exhibits the optimum extraction performance for pyrethroids, while the MM/ZIF$-$8 was the second best and M-MWCNTs was the worst. The possible reason could be that more surface area is made available on the M-MWCNTs after the formation of MM/ZIF$-$8, which is then available for the adsorption of target analytes. Furthermore, the hydrophobic nature and abundance of potential hydrogen bond donors and acceptors of the DES on MM/ZIF$-$8@DES_5 could facilitate its adsorption for pyrethroids. Therefore, MM/ZIF$-$8@DES_5 was chosen as the magnetic adsorbent in the MSPE procedure being developed.

The quantity of the sorbent can obviously affect the extraction performance of the MSPE procedure. To achieve a favorable extraction performance for pyrethroids, 2, 4, 6, 8 and 10 mg of magnetic sorbent were suspended separately to the fortified sample solutions (50 µg L^{-1}). As shown in Figure 3D, 6 mg of adsorbent gives an outstanding extraction performance for pyrethroids and was consequently set as the optimum dosage of the adsorbent.

3.2.3. Effect of Extraction Time

To study the effect of the extraction time on MSPE performance, 2 to 20 min of shaking time was tested. The extraction efficiencies were promoted with the increase of treatment time from 2 to 10 min. After that, the extraction efficiencies were maintained relatively unchanged as the treatment time increased from 10 to 20 min (Figure 3E). Therefore, the adsorption time was set at 10 min.

3.2.4. Effect of pH of Sample Solution

Solution pH is another crucial parameter for the optimization of the MSPE procedure because of its probability of changing the surface charge of the magnetic adsorbent and/or chemical form of the analyte. Therefore, sample solutions with a pH ranging from 4.0 to 8.0 were prepared by adjusting with HCl or NaOH as necessary. As shown in Figure 3F, the extraction performance for six analytes were slightly promoted with the increase of pH value from 4.0 to 6.0, whereas they were reduced when the pH value increased past 7.0. The probable cause was that alkaline conditions affect the stability of pyrethroids. Furthermore, the pH values for most of the tea drinks were approximately 5.6–5.8 [28]. Therefore, the tea drink samples needed no adjustment.

3.2.5. Selection of the Type and Volume of Desorption Solvent

The type and usage of a desorption solvent played a key role in achieving a favorable recovery of the target analytes. To study the affection of eluent on desorption performance, HPLC grade acetonitrile, methanol, acetone, n-hexane, and ethyl acetate were selected as the potential desorption solvents for this study, and the results indicated that the ethyl acetate gave the optimum desorption performance (Figure 3G). All candidate desorption solvents were obtained from Sigma-Aldrich (ST, USA). Meanwhile, the volume of eluent was optimized from 1.5 to 3.5 mL, and the results illustrated that 2.5 mL of ethyl acetate was adequate for the desorption process (Figure 3H). Hence, 2.5 mL of ethyl acetate was selected as the ideal desorption conditions in further studies.

3.3. Method Validation

To confirm the performance of the established method, different parameters of linearity, limit of detection (LOD), and precision were investigated by analyzing fortified blank samples (Table 4). The linear range of the as-developed method was acquired by analyzing working solutions containing six pyrethroids (0.5, 1, 2, 5, 10, 20, 50, 100, 200, and 500 µg L^{-1}) and plotting the concentration versus the peak area. The results suggested good linearity for the six pyrethroids with the correlation coefficient (R^2) ranging from 0.9905 to 0.9925. The LODs were derived from the signal-to-noise ratio (S/N) of 3, and the results are 0.08–0.33 µg L^{-1}. The precision of the developed method was studied by measuring the intra- and inter-day RSDs, and the RSD values were less than 5.58% and 8.58%, respectively. All the above results suggested the established method possesses a prospective performance for the determination of pyrethroid pesticide residues in tea drinks.

Table 4. Analytical parameters of MM/ZIF−8@DES$_5$–MSPE-GC-MS/MS method for the analysis of the six pyrethroids from tea drinks.

Analytes	Calibration	Linear Range (µg L^{-1})	R^2	Intraday RSD (%)	Interday RSD (%)	LOD (µg L^{-1})
Cyhalothrin	y = 3703.2x −46478	0.5–400	0.9905	4.00	7.57	0.08
Cyfluthrin	y = 3791.2x −47015	0.5–400	0.9907	4.84	7.56	0.33
Cypermethrin	y = 8168.8x −92816	0.5–400	0.9925	2.99	5.77	0.22
Flucythrinate	y = 19644.0x −251300	0.5–400	0.9910	3.50	5.83	0.13
Fenvalerate	y = 3037.9x −38641	0.5–400	0.9912	4.83	7.66	0.24
Fluvalinate	y = 5608.7x −59333	0.5–400	0.9922	5.58	8.58	0.10

3.4. Comparison of the Proposed MSPE with Other Published Methods

To demonstrate the potential application of MM/ZIF−8@DES$_5$ as an adsorbent in MSPE, the as-developed method was compared with several published methods (Table 5). After careful consideration of the exhibited parameters, the proposed method showed a similar performance to those previously reported.

Table 5. Comparison of the developed pretreatment technique with other published methods.

Method	Sorbent	Sample Amount (mL)	Sorbent Amount (mg)	Extraction Time (min)	Volume of Eluent (mL)	Linear Range (µg L^{-1})	LOD (µg L^{-1})	Ref.
MSPE-DLLMESFO-GC-ECD	Fe$_3$O$_4$/MIL-101(Cr)	50	10	10	methanol, 0.4	0.05–10	0.008–0.015	[29]
MPSE-GC	Magnetic silica aerogels	2.5	30	10	ethyl acetate	0.04–8	0.008–0.024	[30]
dSPE-UFLC-UV	Fe$_3$O$_4$/C/PANI microbowls	150	8	12	methanol, 3	0.1–20	0.025–0.032	[31]
MSPE-HPLC-UV	Fe$_3$O$_4$-MCNTs	10	40	15	5% acetic acid acetonitrile, 3	0.05–25 µg g^{-1}	0.010–0.018 µg g^{-1}	[32]
MSPE-GC-MS/MS	MM/ZIF−8@DES	5	6	10	ethyl acetate, 3	0.5–400	0.08–0.33	This work

3.5. Real Sample Analysis

Under the optimal conditions, the as-developed method was used to determine six pyrethroid residues in three tea drink samples (red tea, green tea, and oolong tea, purchased from a local market), and negative results of pyrethroid residues were obtained; due to the response values of the MS instrument for analytes, which were lower than the LODs (Table 6). To confirm the adaptability of the proposed method for real tea drink samples, standard solutions at 10 and 100 µg L^{-1} were spiked in real samples and investigated to accompany the real sample analysis. The recovery experiments were performed in triplicate. The recovery results suggested that the established method exhibits a satisfactory performance for the determination of pyrethroid residues in tea drinks.

Table 6. Analytical results for the determination of pyrethroids in real tea drink samples.

| Matrix | Analyte | Spiked Levels (µg L^{-1}, $n = 3$) | | | | |
| | | 0 | 10 | | 100 | |
		Found	Recovery (%)	RSD (%)	Recovery (%)	RSD (%)
Red tea	Cyhalothrin	<LOD	72.5	7.1	85.4	2.5
	Cyfluthrin	<LOD	76.3	3.2	86.1	2.8
	Cypermethrin	<LOD	77.5	6.6	82.5	5.3
	Flucythrinate	<LOD	70.4	6.2	87.7	2.1
	Fenvalerate	<LOD	76.1	3.4	87.2	3.6
	Fluvalinate	<LOD	78.3	4.9	96.4	4.3
Green tea	Cyhalothrin	<LOD	83.7	6.0	86.0	4.1
	Cyfluthrin	<LOD	75.2	5.1	84.0	1.9
	Cypermethrin	<LOD	79.1	4.6	82.2	2.1
	Flucythrinate	<LOD	77.3	6.8	89.3	3.5
	Fenvalerate	<LOD	72.8	8.7	87.4	3.1
	Fluvalinate	<LOD	80.7	7.9	86.3	4.5
Oolong tea	Cyhalothrin	<LOD	74.4	7.1	86.0	2.4
	Cyfluthrin	<LOD	75.2	6.8	83.8	2.9
	Cypermethrin	<LOD	72.3	10.0	88.8	7.6
	Flucythrinate	<LOD	73.6	3.7	93.1	3.9
	Fenvalerate	<LOD	81.1	5.3	91.8	1.9
	Fluvalinate	<LOD	82.6	7.9	86.3	4.5

4. Conclusions

In this study, a novel and efficient DES-type surfactant functionalized MM/ZIF−8 composite (MM/ZIF−8@DES$_5$) was successfully prepared and selected as an effective magnetic adsorbent for the determination of pyrethroids in tea drink samples. Material characterization indicated that the MM/ZIF−8@DES$_5$ possesses a sufficient specific surface area, a decent pore volume, and superparamagnetism, which will enable the rapid separation of pyrethroids from tea drink samples. Validation of the proposed method suggested excellent linearity, low LODs, and good precision. The developed method possesses a considerable future for the monitoring of organic pollutants in the environment or in food samples.

Author Contributions: Conceptualization, Methodology, Writing—review and editing, X.H.; Investigation, Methodology, Writing—original draft, H.L.; Methodology, Data curation, X.X.; Validation, Methodology, G.C.; Methodology, Data curation, L.L.; Validation, Writing—original draft, Y.Z.; Conceptualization, Writing—review and editing, G.L.; Funding acquisition, Project administration, Writing—review and editing, D.X. All authors have read and agreed to the published version of the manuscript.

Funding: This work was supported by the Agricultural Science and Technology Innovation Program of CAAS (CAAS-ZDRW202011, CAAS-TCX2019025-5), the China Agriculture Research System of MOF and MARA (CARS-23-E03), and the National Key Research Development Program of China (2020YFD1000300).

Data Availability Statement: The data presented in this study are available on request from the corresponding author.

Conflicts of Interest: The authors declare that they have no known competing financial interest or personal relationships that could have appeared to influence the work reported in this paper.

References

1. Jiang, H.-L.; Li, N.; Cui, L.; Wang, X.; Zhao, R.-S. Recent application of magnetic solid phase extraction for food safety analysis. *TrAC Trend. Anal. Chem.* **2019**, *120*, 115632. [CrossRef]
2. Keçili, R.; Ghorbani-Bidkorbeh, F.; Dolak, İ.; Canpolat, G.; Karabörk, M.; Hussain, C.M. Functionalized magnetic nanoparticles as powerful sorbents and stationary phases for the extraction and chromatographic applications. *TrAC Trends Anal. Chem.* **2021**, *143*, 116380. [CrossRef]
3. Huang, Y.; Li, Y.; Luo, Q.; Huang, X. One-step preparation of functional groups-rich graphene oxide and carbon nanotubes nanocomposite for efficient magnetic solid phase extraction of glucocorticoids in environmental waters. *Chem. Eng. J.* **2021**, *406*, 126785. [CrossRef]
4. Liu, Z.; Wang, J.; Wang, Z.; Xu, H.; Di, S.; Zhao, H.; Qi, P.; Wang, X. Development of magnetic solid phase extraction using magnetic amphiphilic polymer for sensitive analysis of multi-pesticides residue in honey. *J. Chromatogr. A* **2022**, *1664*, 462789. [CrossRef]
5. Meseguer-Lloret, S.; Torres-Cartas, S.; Gómez-Benito, C.; Herrero-Martínez, J.M. Magnetic molecularly imprinted polymer for the simultaneous selective extraction of phenoxy acid herbicides from environmental water samples. *Talanta* **2022**, *239*, 123082. [CrossRef]
6. Li, T.; Lu, M.; Gao, Y.; Huang, X.; Liu, G.; Xu, D. Double layer MOFs M-ZIF-8@ZIF-67: The adsorption capacity and removal mechanism of fipronil and its metabolites from environmental water and cucumber samples. *J. Adv. Res.* **2020**, *24*, 159–166. [CrossRef]
7. Rathinavel, S.; Priyadharshini, K.; Panda, D. A review on carbon nanotube: An overview of synthesis, properties, functionalization, characterization, and the application. *Mat. Sci. Eng. B* **2021**, *268*, 115095. [CrossRef]
8. Ehzari, H.; Safari, M.; Samimi, M. Signal amplification of novel sandwich-type genosensor via catalytic redox-recycling on platform MWCNTs/Fe$_3$O$_4$@TMU-21 for BRCA1 gene detection. *Talanta* **2021**, *234*, 122698. [CrossRef]
9. Wang, Y.; Zhang, N.; Tan, D.; Qi, Z.; Wu, C. Facile synthesis of enzyme-embedded metal–organic frameworks for size-selective biocatalysis in organic solvent. *Front. Bioeng. Biotechnol.* **2020**, *8*, 714. [CrossRef]
10. Miensah, E.D.; Khan, M.M.; Chen, J.Y.; Zhang, X.M.; Wang, P.; Zhang, Z.X.; Jiao, Y.; Liu, Y.; Yang, Y. Zeolitic imidazolate frameworks and their derived materials for sequestration of radionuclides in the environment: A review. *Crit. Rev. Env. Sci. Tec.* **2020**, *50*, 1874–1934. [CrossRef]
11. Pan, Y.; Liu, Y.; Zeng, G.; Zhao, L.; Lai, Z. Rapid synthesis of zeolitic imidazolate framework-8 (ZIF-8) nanocrystals in an aqueous system. *Chem. Commun.* **2011**, *47*, 2071–2073. [CrossRef] [PubMed]
12. Liu, H.; Jiang, L.; Lu, M.; Liu, G.; Li, T.; Xu, X.; Li, L.; Lin, H.; Lv, J.; Huang, X.; et al. Magnetic solid-phase extraction of pyrethroid pesticides from environmental water samples using deep eutectic solvent-type surfactant modified magnetic zeolitic imidazolate framework-8. *Molecules* **2019**, *24*, 4038. [CrossRef] [PubMed]
13. Smith, E.L.; Abbott, A.P.; Ryder, K.S. Deep eutectic solvents (DESs) and their applications. *Chem. Rev.* **2014**, *114*, 11060–11082. [CrossRef]
14. Herce-Sesa, B.; López-López, J.A.; Moreno, C. Advances in ionic liquids and deep eutectic solvents-based liquid phase microextraction of metals for sample preparation in environmental analytical chemistry. *TrAC Trend. Anal. Chem.* **2021**, *143*, 116398. [CrossRef]
15. Santos, L.B.; Assis, R.S.; Barreto, J.A.; Bezerra, M.A.; Novaes, C.G.; Lemos, V.A. Deep eutectic solvents in liquid-phase microextraction: Contribution to green chemistry. *TrAC Trend. Anal. Chem.* **2022**, *146*, 116478. [CrossRef]
16. Song, X.; Zhang, R.; Xie, T.; Wang, S.; Cao, J. Deep eutectic solvent micro-functionalized graphene assisted dispersive micro solid-phase extraction of pyrethroid insecticides in natural products. *Front. Chem.* **2019**, *7*, 594. [CrossRef] [PubMed]
17. Chrustek, A.; Hołyńska-Iwan, I.; Dziembowska, I.; Bogusiewicz, J.; Wróblewski, M.; Cwynar, A.; Olszewska-Słonina, D. Current research on the safety of pyrethroids used as insecticides. *Medicina* **2018**, *54*, 61. [CrossRef]
18. Bhatt, P.; Bhatt, K.; Huang, Y.; Lin, Z.; Chen, S. Esterase is a powerful tool for the biodegradation of pyrethroid insecticides. *Chemosphere* **2020**, *244*, 125507. [CrossRef]
19. Zhu, Q.; Yang, Y.; Zhong, Y.; Lao, Z.; O'Neill, P.; Hong, D.; Zhang, K.; Zhao, S. Synthesis, insecticidal activity, resistance, photodegradation and toxicity of pyrethroids (A review). *Chemosphere* **2020**, *254*, 126779. [CrossRef]
20. Bradberry, S.M.; Cage, S.A.; Proudfoot, A.T.; Vale, J.A. Poisoning due to pyrethroids. *Toxico. Rev.* **2005**, *24*, 93–106. [CrossRef]
21. Aznar-Alemany, Ò.; Eljarrat, E. Introduction to Pyrethroid Insecticides: Chemical Structures, Properties, Mode of Action and Use. In *Pyrethroid Insecticides*; Eljarrat, E., Ed.; Springer: Berlin/Heidelberg, Germany, 2020; Volume 92, pp. 1–16. [CrossRef]
22. Deng, W.; Yu, L.; Li, X.; Chen, J.; Wang, X.; Deng, Z.; Xiao, Y. Hexafluoroisopropanol-based hydrophobic deep eutectic solvents for dispersive liquid-liquid microextraction of pyrethroids in tea beverages and fruit juices. *Food Chem.* **2019**, *274*, 891–899. [CrossRef] [PubMed]
23. Liu, G.; Li, L.; Huang, X.; Zheng, S.; Xu, X.; Liu, Z.; Zhang, Y.; Wang, J.; Lin, H.; Xu, D. Adsorption and removal of organophosphorus pesticides from environmental water and soil samples by using magnetic multi-walled carbon nanotubes @ organic framework ZIF-8. *J. Mater. Sci.* **2018**, *53*, 10772–10783. [CrossRef]
24. Huang, X.; Liu, Y.; Liu, G.; Li, L.; Xu, X.; Zheng, S.; Xu, D.; Gao, H. Preparation of a magnetic multiwalled carbon nanotube@polydopamine/zeolitic imidazolate framework-8 composite for magnetic solid-phase extraction of triazole fungicides from environmental water samples. *RSC Adv.* **2018**, *8*, 25351–25360. [CrossRef] [PubMed]

25. Huang, Y.; Wang, Y.; Pan, Q.; Wang, Y.; Ding, X.; Xu, K.; Li, N.; Wen, Q. Magnetic graphene oxide modified with choline chloride-based deep eutectic solvent for the solid-phase extraction of protein. *Anal. Chim. Acta* **2015**, *877*, 90–99. [CrossRef]
26. Wang, J.; Liu, X.; Wei, Y. Magnetic solid-phase extraction based on magnetic zeolitic imazolate framework-8 coupled with high performance liquid chromatography for the determination of polymer additives in drinks and foods packed with plastic. *Food Chem.* **2018**, *256*, 358–366. [CrossRef]
27. Zhang, H.; Wang, Y.; Xu, K.; Li, N.; Wen, Q.; Yang, Q.; Zhou, Y. Ternary and binary deep eutectic solvents as a novel extraction medium for protein partitioning. *Anal. Methods* **2016**, *8*, 8196–8207. [CrossRef]
28. Yang, X.; Lin, X.; Mi, Y.; Gao, H.; Li, J.; Zhang, S.; Zhou, W.; Lu, R. Ionic liquid-type surfactant modified attapulgite as a novel and efficient dispersive solid phase material for fast determination of pyrethroids in tea drinks. *J. Chromatogr. B* **2018**, *1089*, 70–77. [CrossRef]
29. Lu, N.; He, X.; Wang, T.; Liu, S.; Hou, X. Magnetic solid-phase extraction using MIL-101(Cr)-based composite combined with dispersive liquid-liquid microextraction based on solidification of a floating organic droplet for the determination of pyrethroids in environmental water and tea samples. *Microchem. J.* **2018**, *137*, 449–455. [CrossRef]
30. Feng, T.; Ye, X.; Zhao, Y.; Zhao, Z.; Hou, S.; Liang, N.; Zhao, L. Magnetic silica aerogels with high efficiency for selective adsorption of pyrethroid insecticides in juices and tea beverages. *New J. Chem.* **2019**, *43*, 5159–5166. [CrossRef]
31. Wang, Y.; Sun, Y.; Gao, Y.; Xu, B.; Wu, Q.; Zhang, H.; Song, D. Determination of five pyrethroids in tea drinks by dispersive solid phase extraction with polyaniline-coated magnetic particles. *Talanta* **2014**, *119*, 268–275. [CrossRef]
32. Gao, L.; Chen, L. Preparation of magnetic carbon nanotubes for separation of pyrethroids from tea samples. *Microchim. Acta* **2013**, *180*, 423–430. [CrossRef]

Review

Recent Advances in and Applications of Electrochemical Sensors Based on Covalent Organic Frameworks for Food Safety Analysis

Hongwei Zhu [1,2,†], Minjie Li [1,3,†], Cuilin Cheng [2,†], Ying Han [2,4], Shiyao Fu [2,4], Ruiling Li [2], Gaofeng Cao [5], Miaomiao Liu [5], Can Cui [1], Jia Liu [1,3,5,*] and Xin Yang [2,4,*]

[1] Beijing Key Laboratory of Nutrition & Health and Food Safety, Beijing Engineering Laboratory of Geriatric Nutrition & Foods, COFCO Nutrition and Health Research Institute Co., Ltd., Beijing 102209, China; zhuhongwei1994@126.com (H.Z.);liminjie@cofco.com (M.L.); cuican@cofco.com (C.C.)

[2] School of Medicine and Health, Harbin Institute of Technology, Harbin 150001, China; ccuilin@hit.edu.cn (C.C.); hany1994hit@163.com (Y.H.); 15164510974@163.com (S.F.); lirl1995@163.com (R.L.)

[3] Internal Trade Food Science Research Institute Co., Ltd., Beijing 102209, China

[4] School of Chemistry and Chemical Engineering, Harbin Institute of Technology, Harbin 150001, China

[5] COFCO Corporation, Beijing 100020, China; caogf@cofco.com (G.C.); liumiaomiao@cofco.com (M.L.)

* Correspondence: liujiajia1@cofco.com (J.L.); yangxin@hit.edu.cn (X.Y.)

† These authors contributed equally to this work.

Abstract: The international community has been paying close attention to the issue of food safety as a matter of public health. The presence of a wide range of contaminants in food poses a significant threat to human health, making it vital to develop detection methods for monitoring these chemical contaminants. Electrochemical sensors using emerging materials have been widely employed to detect food-derived contaminants. Covalent organic frameworks (COFs) have the potential for extensive applications due to their unique structure, high surface area, and tunable pore sizes. The review summarizes and explores recent advances in electrochemical sensors modified with COFs for detecting pesticides, antibiotics, heavy metal ions, and other food contaminants. Furthermore, future challenges and possible solutions will be discussed regarding food safety analysis using COFs.

Keywords: covalent organic frameworks; electrochemical sensors; food safety; pesticides; antibiotics

Citation: Zhu, H.; Li, M.; Cheng, C.; Han, Y.; Fu, S.; Li, R.; Cao, G.; Liu, M.; Cui, C.; Liu, J.; et al. Recent Advances in and Applications of Electrochemical Sensors Based on Covalent Organic Frameworks for Food Safety Analysis. *Foods* **2023**, *12*, 4274. https://doi.org/10.3390/foods12234274

Academic Editor: Seyed-Ahmad Shahidi

Received: 23 October 2023
Revised: 14 November 2023
Accepted: 20 November 2023
Published: 27 November 2023

1. Introduction

The safety of food is of vital importance to the health of people and to the long-term stability of society in general [1]. Food forms the basis of human survival and is essential for maintaining a stable and sustained existence [2]. Food safety is defined by the Food Safety Law of the People's Republic of China as non-toxic, harmless, and meeting the nutritional requirements without causing acute, subacute, or chronic harm to humans. It is noteworthy that the European Union, the United States, and other countries have very similar definitions of food safety, even if they express it in a slightly different manner.

Globally, the top 100 food and beverage companies generated revenues of USD 1.3 trillion in 2019, equivalent to approximately CNY 9.2 trillion [3]. However, with the achievement of economic globalization, food safety issues have become a worldwide issue that has impacted more than just an individual country or region. Currently, food safety is subjected to many challenges due to differences in the natural environment in different countries and regions. It is possible for food to become unavoidably contaminated during its preparation, transportation, and storage, regardless of how rigorous and meticulous the handling procedures are. As a result of the excessive use of veterinary drugs and pesticides [4] and heavy metal ions [5] and the introduction of illegal additives [6], in particular, food can be contaminated in a variety of ways throughout the food chain [7]. In addition to these factors, hazardous food contaminants deserve special attention because even in low concentrations, they are able to cause serious diseases such as cancer [8], and furthermore, fungi that contaminate food such as aspergillus, penicillium, and neotyphodium [9] pose a serious threat to human health and safety. The food safety industry has experienced some extremely detrimental incidents in recent years, including the melamine incident at Sanlu Group in China in 2008, the salmonella-contaminated peanut butter incident at Peanut Corporation of America from 2008 to 2009, the E. coli contamination of bean sprouts in the European Union in 2011, and the contamination of milk powder in 2013 with Clostridium botulinum toxin by Fonterra in New Zealand. Due to these recent food safety incidents, the global society has been paying close attention to this issue, and many countries and regions have adjusted their policies and intensified their supervision on food regulation [10]. In addition to posing significant risks to the health and safety of the general public, these frequent food safety incidents also cause significant losses for the industries that are directly affected by the incidents.

Having a rational and effective approach for food testing is an essential component of food safety management. Conducting well-informed research on testing techniques can provide powerful assurances regarding food safety being maintained continuously.

1.1. Electrochemical Sensors and Their Role in Food Safety Analysis

The foundation of any food safety program is improved food safety testing techniques, which are key in addressing food safety. Food safety testing methods can be classified as traditional or rapid detection. Usually, conventional methods consist of using techniques such as gas chromatography–mass spectrometry [11], high-performance liquid chromatography [12], and liquid chromatography–mass spectrometry [13,14] to determine the identity of food. These methods are usually performed in laboratories with sophisticated equipment. They are frequently used as reference standards to ensure food safety because of their high sensitivity, accuracy, precision, and repeatability. However, the length of their analysis cycle and their low throughput are their limitations. Rapid detection methods, on the other hand, deliver faster results. It is common to use these methods of qualitative or semi-quantitative screening of target analytes [15]. The benefits of electrochemical detection methods are many, including affordability, simplicity, ease of operation, miniaturization, and diversification, over traditional methods such as spectroscopy and chromatography [16]. In addition, electrochemical detection is suitable for automated control as well as online sensitive and rapid analysis since it can be conducted remotely [17]. They can be applied to biomedical sciences, pharmaceuticals, environmental sciences, and food sciences, and are considered to be one of the most dynamic and promising analytical techniques [18].

The primary objective of electrochemical detection techniques is to qualitatively or quantitatively analyze and measure target substances based on their electrical and electrochemical properties through the use of electrochemical sensors [19]. There are two main components of electrochemical sensors: a molecular recognition system and a system for converting information into electrical signals (the principle of electrochemical sensors is illustrated in Figure 1). Based on the measured chemical parameters, response signals are generated in the form of voltages, currents, or changes in light intensity. These signals are then amplified, converted, and finally transformed into analyzable signals that indicate the amount of target analyte present in the sample using electronic systems [20]. It is widely known that sensors with electrochemical integration are widely used in a variety of fields, including industry, transportation, environmental monitoring, and medical surveillance. Sensors based on electrochemical reactions play an important role in combining sensing technology and electrochemical analysis technology. Electrochemical sensors have been widely applied and developed since the 1960s, with electrodes serving as the basic component [21]. The electrodes play an integral role in the overall performance of electrochemical sensors due to their functionality and interfacial performance. However, one of the main challenges is making electrodes more responsive and selective to desired reactions. Nanotechnology has made rapid progress since the 1980s, resulting in many nanomaterials with exceptional performance and unique structures, and these materials have excellent biocompatibility, a high surface chemical activity, a large specific surface area, and a high electron transfer efficiency, thus facilitating the use of nanomaterials (for instance, COF [22–24], MOF [25,26], MIP [27–29], among others [30,31]) in electrochemical sensing). New types of electrochemical sensors have been developed as a result of the convergence of nanotechnology and sensing technology, which has attracted increasing attention. In addition to providing the rapid identification of basic food components, electrochemical sensors are capable of detecting harmful substances such as heavy metal ions [32], foodborne pathogens [33], pesticide residues [34], and food additives [35].

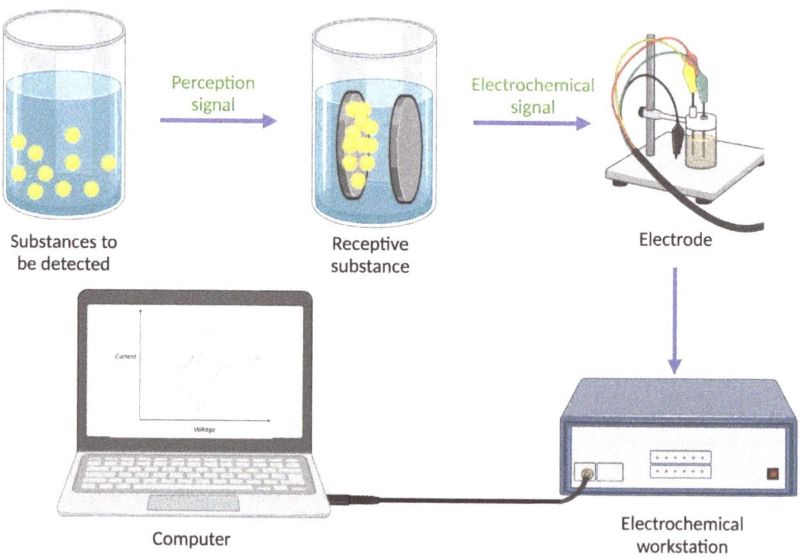

Figure 1. Diagram of an electrochemical sensor. (The mechanism map was created with BioRender.com.)

1.2. Covalent Organic Frameworks (COFs) and Their Potential Applications in Sensor Technology

As a general rule, nanomaterials are materials with at least one dimension out of the three dimensions within the nanometer size range (1–100 nm), or they are formed from basic constituents with such dimensions. Nanomaterials have unique physicochemical properties in optics, electronics, magnetism, heat, mechanics, and other fields as a result of their unique surface effects, small size effects, quantum effects, and macroscopic quantum tunneling effects [36]. The use of these technologies is widespread in fields such as food technology, electronics manufacturing, chemical engineering, and many others [37–39]. Moreover, nanomaterials have a small size effect, leading to a large specific surface area and a high surface energy, and they have abundant surface-active sites and exhibit an ease of functionalization. This contributes to a high catalytic efficiency as well as an excellent biocompatibility, greatly enhancing their potential of being electrochemically research [40,41]. Furthermore, nanocomposites are composed of materials in which nanoparticles are uniformly dispersed within the matrix material. It is important to note that unlike traditional single-phase nanomaterials, nanocomposites can consist of a combination of metal nanoparticles with resins or gels, polymer materials, porous inorganic materials, porous organic materials, and various types of metal nanoparticles. There have been several developments in nanomaterials so far, including carbon materials (graphene, carbon nanotubes, carbon foam, carbon fibers, carbon spheres, porous carbon materials, etc.), metal–organic frameworks (MOFs), zeolitic imidazolate frameworks (ZIFs), covalent organic frameworks (COFs), among others [42–45].

The covalent organic frameworks (COFs) represent a new class of organic porous materials [46]. In 2005, Yaghi and colleagues were successful in synthesizing two-dimensional

COFs, COF-1 and COF-5, via the condensation reaction between phenylboronic acid and 2,3,6,7,10,11-hexahydroxytriphenylene. These COFs with high surface areas (711 and 1590 m^2 g^{-1}, respectively), high thermal stability, and permanent porosity were compared [47]. Following this, Yaghi proposed three-dimensional COFs in 2007, including COF-102, COF-105, and COF-108 [48]. There has been a great deal of interest in COFs since their introduction, and they have been applied in a variety of fields. As a result of their flexible polygonal frameworks that are easy to design and control, COFs have been widely applied for a variety of purposes, such as catalysis, energy storage, water treatment, drug delivery, among others [49]. Compared to conventional electronic components, COF and overoxidized PEDOT or PEDOT/PSS have better electrical signal transduction. Controlling the electrical properties of overoxidized PEDOT and PEDOT/PSS is the primary method of actuating the device. A change in the electrical conductivity of PEDOT/PSS can be achieved by applying an electric field or conducting an electrochemical reaction, thus allowing the material to be controlled in terms of its properties and functions [50]. In contrast, the COF undergoes physical or chemical changes through the adsorption or desorption of internal molecules, which cause changes in the signaling pathway. It is estimated that over 10,000 papers (from WOS) have been published over the past five years due to the wide potential applications of this unique material across many fields.

2. Fundamentals of COF-Based Electrochemical Sensors

2.1. COFs with Different Chemical Structure Types

It is well known that covalent organic frameworks (COFs) are porous organic materials that are constructed by self-assembling materials linked together by covalent bonds [51,52]. Therefore, they possess unmatched biocompatibility and chemical stability, as well as high surface areas, high porosities, and ease of functionalization, similar to metal–organic frameworks (MOFs) and zeolitic imidazolate frameworks (ZIFs). The highly ordered π-π conjugated system in COFs and their independently accessible regular pores provide high levels of electronic conductivity. It is for this reason that these materials are often used as excellent photocatalysts, for gas adsorption and separation, electrochemical sensing, and energy storage applications [53,54]. The following subsections provide an overview of the different chemical structures of COFs.

2.1.1. The B-O Structure of COFs

In 2005, Yaghi et al. synthesized COF-1 and COF-5, typical examples of the B-O structure [47] (Figure 2A). One method for the synthesis of COF-1 is based on the self-condensation of phenylboronic acid, a process in which the boronic acid molecules in phenylboronic acid undergo dehydration in order to form a two-dimensional B_3O_3 ring (boroxine ring). The boronic acid molecules in 1,4-phenyldiboronic acid have the same capability of undergoing condensation during dehydration to form a layered hexagonal framework (COF-1). It is possible to obtain an extendable layered structure (COF-5) by dehydrating and condensing 1,4-phenyldiboronic acid with 2,3,6,7,10,11-hexahydroxytriphenylene. However, it should be noted that these types of COFs have a poor water stability due to the reversible reaction of boronic acid ester formation, which causes their hydrolysis when exposed to acid, alkali, or atmospheric water vapor, thus impairing the quality of their framework [55].

2.1.2. The Imine Structure of COFs

COFs are typically connected by imine bonds when amines and aldehydes undergo the Schiff base reaction, which produce C=N bonds [56] (Figure 2B). Yaghi et al., in 2009, reported the first example of this type of COF. During the experimental process, COF-300 was found to be structurally stable at 490 °C and insoluble in both water and common organic solvents [57]. The application prospects of COFs connected by imine bonds are greater than those of COFs linked by B-O bonds. In addition, COFs with oxime bonds can be considered as another type of imine-based COFs, which are formed by reacting hydrazine compounds with aldehydes or ketones, such as COF-42 and COF-43 [58].

The Banerjee group introduced functional groups -OH into their structures in order to improve their stability and crystallinity. As a result of the reaction between 2,5-dimethoxybenzaldehyde (Dma) and 2,5-dihydroxybenzaldehyde (Dha) with 5,10,15,20-tetra(4-aminophenyl)-21H,23H-porphine (Tph), the COFs of interest were synthesized. The results indicated that, compared to DmaTph, the O-H···N=C interaction in the DhaTph structure partially protected the COFs from hydrolysis under aqueous and acidic conditions, thereby improving their crystallinity and porosity [59].

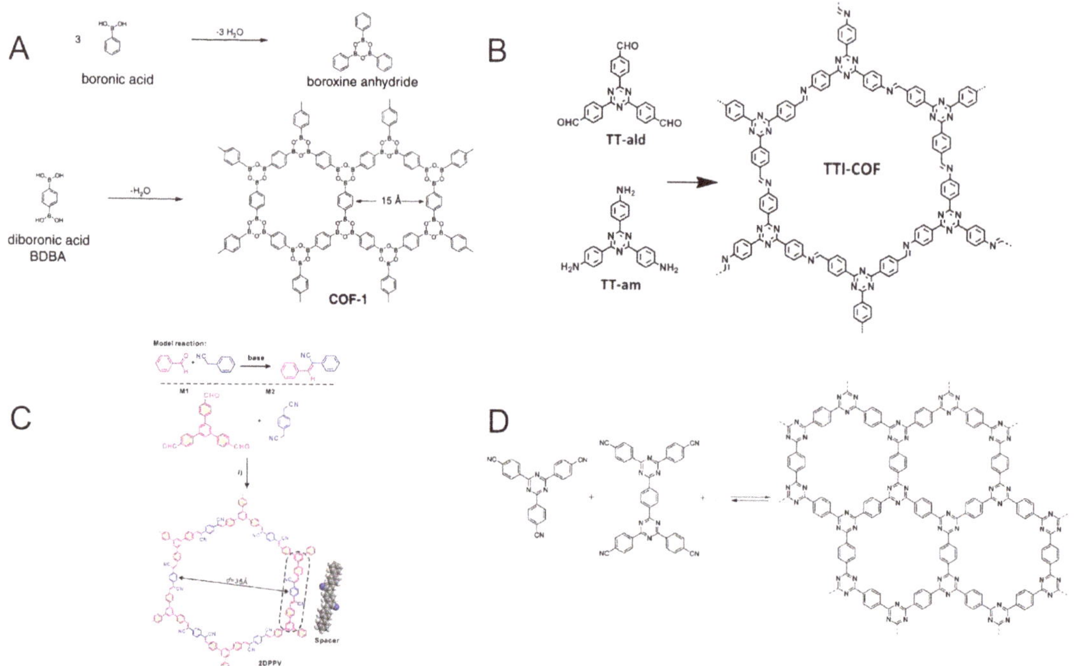

Figure 2. Schematic representation of COF materials synthesized with different chemical structures: (**A**) the B-O structure, Ref. [47], Copyright 2005, The American Association for the Advancement of Science; (**B**) the imine structure, Ref. [56], Copyright 2016, John Wiley and Sons; (**C**) the C=C structure, Ref. [60], Copyright 2016, The Royal Society of Chemistry; and (**D**) the triazine structure, Ref. [61], Copyright 2009, American Chemical Society.

2.1.3. The C=C Structure of COFs

During the Knoevenagel condensation reaction, the active methylene group in a compound is dehydrated and condensed with an aldehyde or ketone under the catalysis of a base, leading to the formation of a thermally stable compound. In spite of this, due to the limitations of the reaction conditions, it has been difficult to apply this principle to the synthesis of COFs for a long time. The first time this principle was applied was in 2016, when Zhang and Feng synthesized two-dimensional conjugated COFs (2DPPV) with C=C connectivity [60]. Figure 2C illustrates the reaction process, in which diphenyl dinitrile and 1,3,5-tris(4-formylphenyl)benzene are used as monomers to produce a two-dimensional-layered framework. By activating it further, carbon nanosheets can be formed, increasing their surface area from 472 m^2 g^{-1} to 880 m^2 g^{-1}. In the field of electrochemistry, it has been successfully applied as a capacitor and catalyst, demonstrating its great potential.

2.1.4. The Triazine Structure of COFs

Covalent triazine-based frameworks (CTFs), which are triazine-based COFs, were first synthesized in 2008 by Kuhn et al. The self-polymerization of dinitrile occurs under the catalysis of zinc chloride at 400 $^\circ$C, followed by the polymerization into polymers based on the triazine structure [61] (Figure 2D). It is important to note that these COFs have excellent thermal stability as well as chemical stability, but because of the high temperatures and strong acid catalysis required for the reaction process, their subsequent applications are limited. There are also many research groups that are dedicated to developing milder preparation methods.

2.1.5. The Other Structure of COFs

Furthermore, there are many other ways of connecting COFs besides those mentioned above, such as C-C [62], aminal [63], imide [64], ester [65], and quinoline [66]. Their chemical stabilities, thermal stabilities, large surface areas, and designable pore sizes make them highly promising in a wide range of applications.

Due to their high functionality, as well as their highly ordered π-π conjugated systems, independent open pores, and high specific surface area, COF materials facilitate rapid electron transfer and energy storage. Moreover, COFs possess electrodes with high specific surface areas and a dense exposure of catalytic active sites, and the interconnected pores facilitate diffusion and contact between the analytes and the active sites. Therefore, it has been found that electrodes constructed using COFs directly or with electrochemically active molecules are ideal electrodes for electrochemical sensing analysis [22,67]. It has been reported that various types of electrochemically active COFs have been developed as a result of COFs' ability to be easily controlled by functional groups. In Wang's research team, an electrochemically active two-dimensional COF$_{\text{Thi-TFPB}}$ was synthesized by introducing sulfur as an electroactive monomer, which was then grown on carbon nanotube surfaces functionalized with amines. It was applied to the construction of ascorbic acid (AA) and pH sensors [68]. In Lu's research laboratory, a topological skeleton COF-LZU1 based on Fe^{3+} coordination was prepared, as well as Fe_3O_4/N composites for enzyme-free plasma component detection [69]. According to Wang's research group, the self-redox-active COF$_{\text{DHTA-TTA}}$ was used as an electroactive material in the construction of electrochemical sensors for H_2O_2, pH, glucose, etc., which demonstrated excellent stability and performance in the detection of these targets [70]. Zhang's research group utilized COF nanocomposites doped with Au NPs as signal probes for catechin testing [71]. It is important to recognize that the output of electrical signals is a critical component in the design of electrochemical sensors. It follows that the crucial issue in the application of COFs to electrochemical sensing is the development of more versatile electrode materials, the design of electroactive COFs, or the development of COFs that are capable of performing more than one function. Consequently, COF materials may be developed and applied to electrochemical analysis with some potential and feasibility.

2.2. Principles of Electrochemical Detection and COF-Based Electrochemical Sensors

The electrochemical sensor detects and quantifies chemical components in a sample using electrochemical principles. The selection of electrode materials is essential for the construction of the electrochemical sensing interface, and COFs have gained considerable attention as highly promising electrode materials. It is well known that COFs possess a variety of porous structures, low toxicity, and excellent biocompatibility, which make them ideal for the construction of sensing interfaces. The application of COFs to electrochemical sensors is therefore becoming increasingly popular. There have been more than 100 publications in this field (from WOS) over the last five years.

It is also possible to modify COFs with different functional groups or metal ions to develop a number of highly specific and targeted sensors [72]. In addition to their outstanding stability, they are widely used in electrochemical sensors due to their high durability [73]. With electrochemical sensors based on COFs, the real-time monitoring of analytes is possible with minimal sample preparation and rapid analysis. Over the last several years, COFs have attracted an increasing amount of attention owing to their excellent performance and their ability to be used in the development of new electrochemical sensors. There has been a steady increase in the number of articles related to COF-based electrochemical sensors since Wang and colleagues detected Pb^{2+} using COF-based electrochemical sensors in 2018 [74]. It is primarily in the food industry that electrochemical sensors based on COF are used for the detection of hazards associated with food.

3. Recent Advances in COF-Based Electrochemical Sensors for Food Safety Analysis

In the field of electrochemical sensing, COFs have been successfully applied due to their captivating structure and properties [73]. It is currently possible to detect a wide range of food contaminants using electrochemical sensors based on COFs. COFs are advantageous because they provides a large number of binding sites and π-π stacking interactions, which speed up charge transfer and enhance the electrochemical performance of the sensor [75]. Consequently, they exhibit a high degree of selectivity, a high sensitivity, and a rapid response time. Throughout this review, COF-based electrochemical sensors are presented and discussed as a means of detecting various food hazards, including pesticides, heavy metal ions, antibiotics, and other relevant substances (Figure 3 and Table 1).

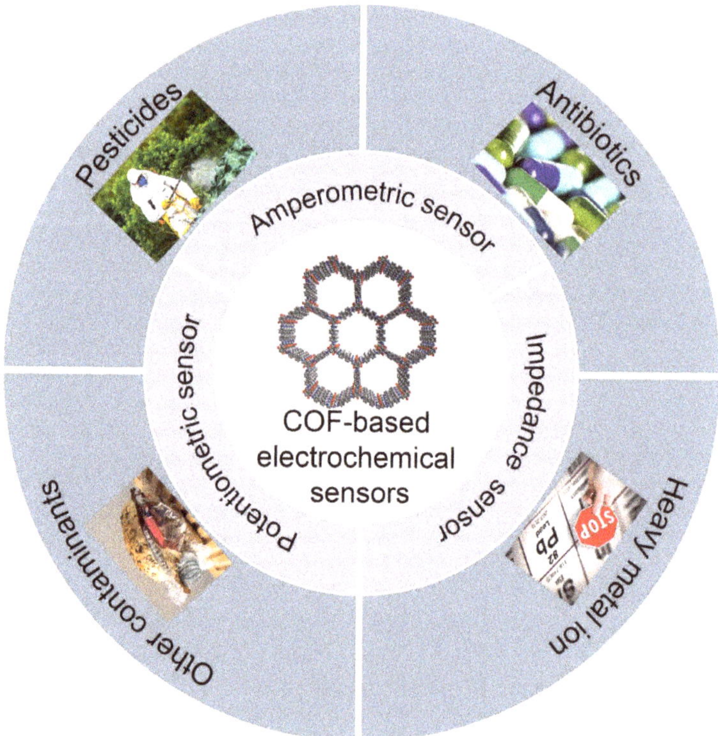

Figure 3. COF-based electrochemical sensing platforms in food safety applications.

3.1. Detection of Pesticides

Pesticides play an important role in agricultural production; however, the residues they leave behind and their degradation products can pose serious threats to ecosystems and human health. It is possible for them to disrupt the ecological balance as well as the major functions of the human body, including the immune system, nervous system, and endocrine system, which can lead to a variety of diseases. There has been a growing concern regarding rapid detection techniques for detecting pesticides in a timely and accurate manner.

It has been reported that some pesticides inhibit enzymes that catalyze substrates, resulting in changes in signal levels and the indirect measurements of pesticide levels [76]. As a consequence of this approach, factors such as enzyme loading and activity have a considerable impact on the performance of the sensor. An ideal platform for enzyme immobilization and protection of enzyme activity is COF, with its large specific surface area and adjustable pore size. Through an amine–aldehyde condensation reaction, Chen et al. constituted a COF that is rich in C=O, NH, and OH groups [77]. The biocatalytic activity of acetylcholinesterase was greatly enhanced when it was immobilized on paper electrodes with COF (Figure 4). This biosensor had a linear range of 0.48–35 μmol/L, with a limit of detection (LOD) of 0.16 μmol/L. Using this electrochemical biosensor, sevin from lettuce juice samples has also been detected. Furthermore, Wang et al. developed an electrochemical sensor for the detection of O,O-dimethyl-O-2,2-dichloroethenyl phosphorothioate (DDVP) by modifying an electrode with ethylene-based electroactive $COF_{Tab-Dva}$ nanofibers (as carriers and conductors) [78]. By interacting with the ethylene groups in COF and the thiol groups in choline thiocholine, the ethylene groups in COF appear to be enriched on the electrode surface, thus improving the sensitivity of the electrode. The current response of the probe is altered due to the formation of repulsion with positively charged choline thiocholine and $[Ru(bpy)_3]^{2+}$. As a result, a low-potential pesticide detection can be achieved. Due to the reduction in the amount of thiol choline catalyzed by AChE as the concentration of pesticides increases, choline thiocholine is less repelled by $[Ru(bpy)_3]^{2+}$, resulting in the generation of redox current signals at the electrode surface. Providing enzymes with a microenvironment of superior chemical stability ensures that they will maintain a higher level of activity regardless of adverse external conditions. The research group of Lu [79] also synthesized COF that contained a large amount of carbonyl groups and used it for the construction of an electrochemical sensor capable of detecting para-hydroxybenzoate in cucumber samples. In addition, Song [80] and Wang [81] independently constructed electrochemical sensors to detect malathion and diazinon on the basis of COFs.

It is noteworthy that, despite the fact that there are few reports regarding the application of COF-based electrochemical sensors for the determination of pesticide residues, these reports demonstrate the promise of these sensors as pesticide analysis tools in the future. For these reasons, more research is urgently required in order to expand the types of COF composites and pesticides that can be evaluated.

3.2. Detection of Heavy Metal Ions

Heavy metal ions, such as mercury, lead, and cadmium, are widespread pollutants found in food. In recent years, heavy metal pollution has become a serious food safety concern due to the development of industries. Through numerous pathways in the food chain, these metals can enter the human body, causing chronic poisoning, neurological disorders, and even cancer [82]. Therefore, ensuring the safety of foods requires the detection of heavy metals. Although various analytical methods have been applied for the qualitative and quantitative analysis of heavy metals, such as atomic fluorescence spectroscopy [83], atomic absorption spectroscopy [84], and inductively coupled plasma mass spectrometry [85], these methods are often complex and expensive, thus limiting their application scope. It is therefore essential to continue to focus on developing rapid and sensitive detection methods.

Zhu and colleagues prepared a highly crystalline COF through the Schiff base reaction between triazine trinitrile (TPA) and 2,4,6-triformylphloroglucinol (TDBA) [86]. The result was the development of an electrochemical sensor based on $COF_{TDBA-TPA}$ capable of simultaneously detecting Cd^{2+}, Cu^{2+}, Pb^{2+}, Hg^{2+}, and Zn^{2+} in drinking water. It was found that Cd^{2+}, Cu^{2+}, Pb^{2+}, Hg^{2+}, and Zn^{2+} had detection limits of 0.922 nM, 0.450 nM, 0.309 nM, 0.208 nM, and 0.526 nM, respectively. Furthermore, $COF_{TDBA-TPA}$ was found to be capable of adsorbing Cd^{2+}, Pb^{2+}, Cu^{2+}, and Hg^{2+}. Using the Schiff base reaction of 2,5,8-triamino-sheptazine (MELE) and 4,4′-(phenyl-(c)[1,2,5]-thiadiazol-4,7-diyl)bisbenzaldehyde (BTDD), they synthesized a COF with multiple metal ion adsorption sites [87]. A $COF_{MELE\text{-}BTDD}$ experiment was performed to detect Cd^{2+}, Pb^{2+}, Cu^{2+}, and Hg^{2+}, with LODs of 0.00474 µM, 0.00123 µM, 0.00114 µM, and 0.00107 µM, respectively.

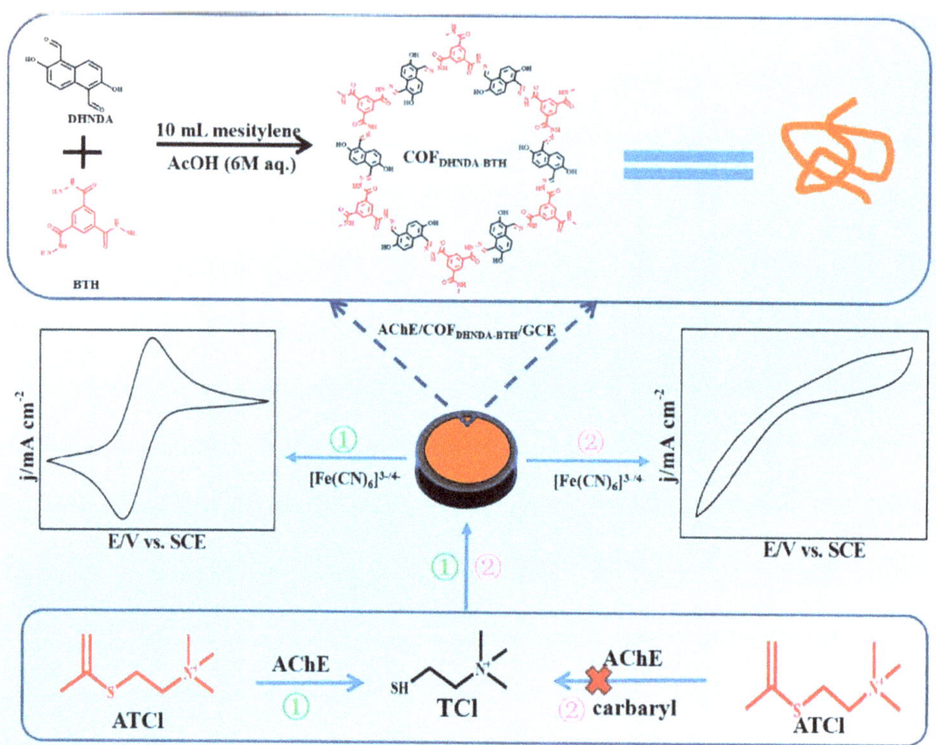

Figure 4. Process of preparing and utilizing the electrochemical sensor for $AChE/COF_{DHNDA\text{-}BTH}$, Ref. [77], Copyright 2022, MDPI.

A triazine-COF modified glassy carbon electrode based on Madrakian's and colleagues' work has been developed as a novel, simple, sensitive, and fast electrochemical sensor for the simultaneous detection of Pb^{2+} and Hg^{2+} [88]. In terms of Pb^{2+} and Hg^{2+}, the linear range was 0.01–0.3 µmol/L, and their lowest detectable limits were 0.72×10^{-3} and 1.2×10^{-2} µmol/L, respectively. Moreover, the detection of Pb^{2+} and Hg^{2+} in food samples was conducted using an electrochemical sensor. A novel glassy carbon electrode was proposed by Madrakian et al. in which a bismuth film, triazine-COF nano-composite materials, and Fe_3O_4 nanoparticles were incorporated into the electrode [89]. Glassy carbon electrodes were capable of selectively detecting Pb^{2+} with a limit of detection (LOD) of 0.95 nmol/L. Based on intercalated composite materials, Zhu and his colleagues developed an electrochemical sensor for the detection of heavy metal ions. Based on the scheme presented in Figure 5, COF-V was synthesized by reacting 1,3,5-tris(4-aminophenyl) benzene with 2,5-divinylterephthalaldehyde [72]. The reaction of AIBN, trithiocyanuric acid, and COF-V resulted in the preparation of COF-SH. Graphene and COF-SH were intercalated onto a glassy carbon electrode (GCE) to produce the intercalated composites. As a result of the good enrichment effect of COF-SH on heavy metal ions and the superior conductivity of graphene, the electrochemical sensor demonstrated excellent performances in the detection of heavy metal ions. Cd^{2+}, Pb^{2+}, Cu^{2+}, and Hg^{2+} each have a detection limit of 0.3, 0.2, 0.2, and 1.1 µg/L, respectively.

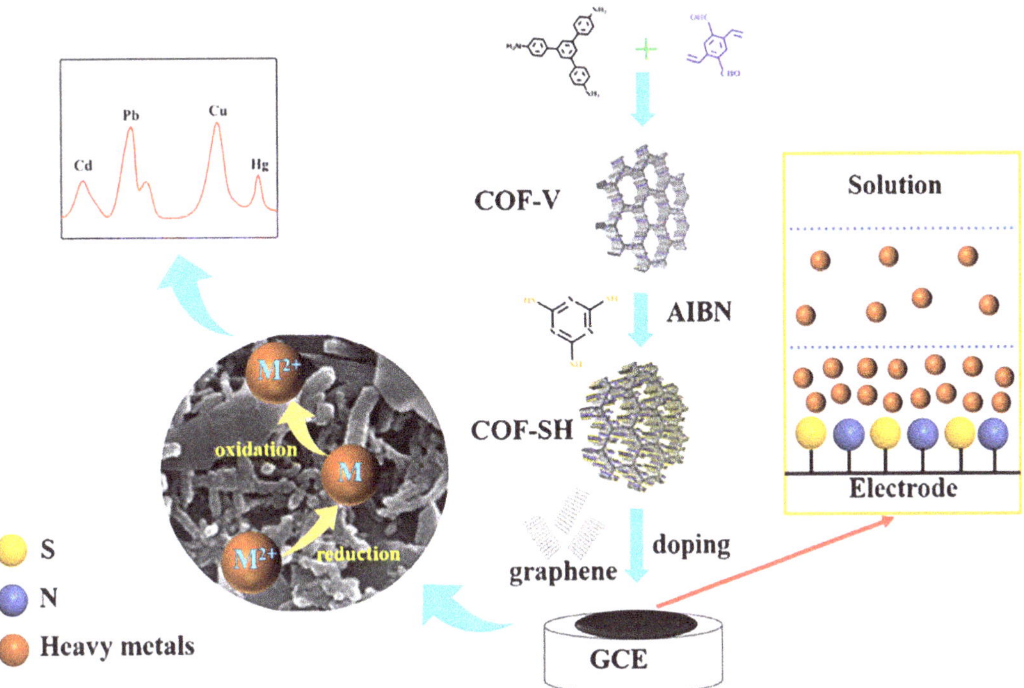

Figure 5. Process of preparing and utilizing the electrochemical sensor for G/COF-SH/GCE, Ref. [72], Copyright 2021, Springer Nature.

3.3. Detection of Antibiotics

It has been demonstrated that antibiotics are effective for inhibiting or killing pathogens; thus, they are widely used in the prevention and treatment of diseases caused by bacteria, fungi, molds, or other microorganisms [90] Antibiotics have become more widely used in the livestock industry as a result of the continued expansion of the industry. It is becoming increasingly important to address the issue of antibiotic residues in this case. There are a number of health problems that can be caused by excessive residual antibiotics, including abnormal blood levels, liver toxicity, and allergic reactions [91]. Research has utilized COFs as scaffolds for immobilizing recognition elements or enhancing electrochemical performance using electroactive COFs for antibiotic detection using COF-based electrochemical sensors. There is no doubt that electrochemical sensors based on COF are highly sensitive, stable, and interference resistant. Nonetheless, the layered structure of 2D COFs allows for an easy encapsulation of their internal active sites, which may restrict the transfer of electrons. It is also essential to consider coating uniformity in order to ensure that modified electrodes perform as expected.

The use of quinolone antibiotics in the treatment and prevention of diseases in humans and animals is widespread. There may be negative environmental and health effects associated with the excessive use of antibiotics. As reported by Du et al., Py-M-COF is synthesized by the condensation reaction of 1,3,6,8-tetra(4-formylphenyl)pyrene (TF-PPy) with cyanuric triamide [23]. The results of electrochemical impedance spectroscopy (ESI) measurements indicated that the Py-M-COF electrochemical sensor could detect enrofloxacin (ENR) and ampicillin (AMP) with extreme sensitivity. The linear response range was 0.12–2000 pg/mL for ENR and 0.001–1000 pg/mL for AMP, respectively, with the lowest detection limits of 6.07 fg/mL and 0.04 fg/mL. It was found that the COF-based sensing system had a higher sensitivity than graphitic carbon nitride (g-C$_3$N$_4$) and amino-functionalized graphene oxide (GO-NH$_2$). COF contains a π-conjugated framework, which provides a higher charge carrier mobility for signals and additional anchoring points for aptamers. It has been demonstrated that Pan et al. prepared TAPB-PDA-COFs/AuNPs by in situ embedding of Au nanoparticles within TAPB-PDA-COFs (formed by the Schiff base reaction of 1,3,5-tri(4-aminophenyl)benzene (TAPB) and p-phenylenedialdehyde (PDA)) [92]. TAPB-PDA-COFs/AuNPs/GCE exhibited a high performance in ENR determination; the results demonstrated that ENR had two linear response ranges between 0.05–10 μM and 10–120 μM, with a LOD of 0.041 μM.

The COF@NH$_2$-CNT composite material was prepared by Sun et al. [67]. to detect furazolidone (NF), by taking advantage of the high surface area of COFs and the excellent conductivity of NH$_2$-CNTs. Due to COFs' efficient adsorption capacity for furazolidone, the sensor was highly sensitive and responded rapidly. COFs were further applied to electrochemical sensors through this strategy [67]. Through the combination of COF$_{TFPB-DHzDS}$, Pt NPs, and rGO, Du and his colleagues developed an electrochemical sensor for the sensitive determination of furazolidone [93]. TFPB and 2,5-bis(3-(ethylthio)propoxy)benzaldehyde hydrazone (DHzDS) were reacted via the Schiff base reaction to produce COF$_{TFPB-DHzDS}$, which was then grown on the surface of rGO-NH2. An in situ reduction method was used to load the Pt NPs onto the COF$_{TFPB-DHzDS}$@rGO. Despite the low detection limit of 0.23 μM and a wide linear range of 0.69 μM to 110 μM, the paper-based electrochemical sensor had a high level of sensitivity.

Antibiotics classified as sulfonamides are broad-spectrum antibiotics used exclusively for treating infections caused by bacteria. Xu et al. developed an electrochemical sensor that is capable of detecting sulfonamide drugs (SMRs) (Figure 6) [94]. MIP/MoS$_2$/NH$_2$-MWCNT@COF/GCE was produced by coating GCE with NH$_2$-MWCNT@COF and MoS$_2$ nanosheets, followed by electrochemical polymerization to obtain MIP/MoS$_2$/NH$_2$-MWCNT@COF/GCE. It was found that the electrochemical sensor prepared for SMR showed a broad response range, from 3.0×10^{-7} M to 2.0×10^{-4} M, with the lowest detection limit of 1.1×10^{-7} M.

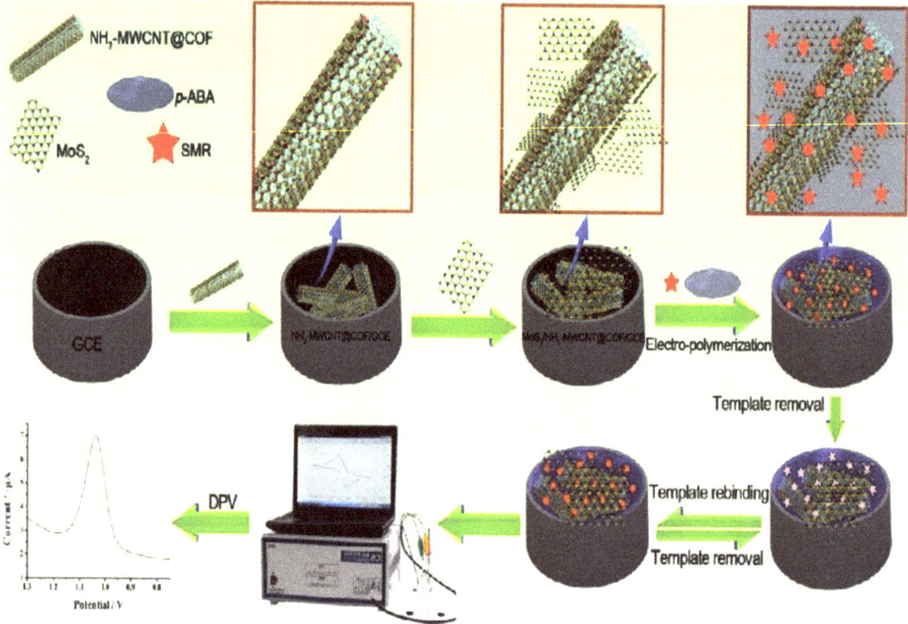

Figure 6. Process of preparing and utilizing the electrochemical sensor for MIP/MoS$_2$/NH$_2$-MWCNT@COF/GCE, Ref. [94], Copyright 2019, Elsevier B.V.

It is commonly known that tetracycline (TC) is a type of antibiotic that can lead to drug resistance and other side effects, such as allergic reactions, kidney toxicity, and liver damage. The detection of tetracycline antibiotics is currently performed using electrochemical sensors based on COFs in order to further improve their stability and portability. A portable on-site electrochemical sensor similar to that proposed by Yukun Yang et al. [95] based on the use of surface molecularly imprinted polymers (MIPs) modified with magnetic COFs (Fe_3O_4@COFs@MIPs) for the sensitive and rapid determination of TC has been proposed. TC is detectable at concentrations from 1×10^{-10} to 1×10^{-4} g/mL, with a limit of detection (LOD) of 2.4×10^{-11} g/mL. Milk and chicken samples have also been successfully tested using the prepared sensor.

3.4. Detection of Other Contaminants

Other contaminants, such as illegal additives, are also threats to the safety of food, in addition to the previously mentioned pollutants. It is possible to selectively detect these targets through the design of COF-based sensing approaches in order to address these challenges.

There is no doubt that tertiary butylhydroquinone (TBHQ), a good antioxidant, plays a critical role in the prevention of lipid oxidation, but the high doses of TBHQ may cause carcinogenesis [96]. Using Co_3O_4@TAPBDMTP-COF as the sensor substrate, Chen et al. were able to more easily and rapidly detect TBHQ, owing to the excellent electrocatalytic property and the large surface area of COFs [97]. Compared with other methods, this approach exhibits higher sensitivity and selectivity towards TBHQ, with a limit of detection as low as 0.02 µM, and it can effectively detect the lower levels of TBHQ present in edible oil samples. There are many adverse effects associated with bisphenols, which are commonly found in plastic food packaging materials [98]. In their study of bisphenol BPS and bisphenol A, Qiao et al., developed a ratio electrochemical sensor able to measure both compounds simultaneously. As a result of the modification of carbon cloth electrodes with silver nanoparticles (COF/AgNPs/CC), this ratio sensor exhibits an excellent electrocatalytic activity toward both bisphenol A and bisphenol BPS, demonstrating a large amount of electrocatalytic surface area and good conductivity [99]. There is no difference between the detection limits for bisphenol A and bisphenol B at 0.15 µmol/L.

Table 1. Statistical analysis of the performance of electrochemical sensors based on different coefficients of friction.

Electrode	Target Substance	Detection Methods	LOD	Linear Range	References
AChE/COF$_{DHNDA-BTH}$/GCE	Carbaryl	Cyclic voltammetry	0.16 µmol/L	0.48–35 µmol/L	[77]
AChE/COF$_{Tab-Dva}$/GCE	DDVP	Cyclic voltammetry	0.11 µM	0.33–30 µM	[78]
GC/COF1/AChE/GCE	Paraoxon	Cyclic voltammetry	1.4 ng/mL	10–1000 ng/mL	[79]
AChE/COF-LZU1/3D-KSC	Trichlorfon	Differential pulse voltammetry	0.067 ng/mL	0.2–19 ng/mL	[80]
COF@MWCNTs	Malathion	Differential pulse voltammetry	0.5 nM	1–10 nM	[81]
COF$_{TDBA-TPA}$/GCE	Cd^{2+}	Square wave anodic stripping voltammetry	0.922 nM	2.8–8000 nM	[86]
	Pb^{2+}		0.309 nM	0.939–4000 nM	
	Cu^{2+}		0.45 nM	1.36–8000 nM	
	Hg^{2+}		0.208 nM	0.632–8000 nM	
	Zn^{2+}		0.526 nM	1.41–7000 nM	
COF$_{MELE-BTDD}$/GCE	Cd^{2+}	Square wave anodic stripping voltammetry	4.74 nM	14.2–4000 nM	[87]
	Pb^{2+}		1.23 nM	3.7–4000 nM	
	Cu^{2+}		1.14 nM	3.4–4000 nM	
	Hg^{2+}		1.07 nM	3.2–4000 nM	
SNW1/GCE	Pb^{2+}	Anodic stripping square wave voltammetry	0.00072 µmol/L	0.01–0.3 µmol/L	[88]
	Hg^{2+}		0.01211 µmol/L	0.05–0.3 µmol/L	
Fe$_3$O$_4$@SNW1/GCE	Pb^{2+}	Square wave anodic stripping voltammetry	0.95×10^{-3} µmol/L	0.003–0.3 µmol/L	[89]
G/COF-SH/GCE	Cd^{2+}	Square wave voltammetry	0.3 µg/L	1–1000 µg/L	[72]
	Pb^{2+}		0.2 µg/L	1–800 µg/L	
	Cu^{2+}		0.2 µg/L	1–800 µg/L	
	Hg^{2+}		0.1 µg/L	5–1000 µg/L	
AptENR/Py-M-COF/AE	Enrofloxacin	Electrochemical impedance spectroscopy	6.07 fg/mL	0.01−2000 pg/mL	[23]
AptAMP/Py-M-COF/AE	Ampicillin		0.04 fg/mL	0.001−1000 pg/mL	
TAPB-PDA-COFs/AuNPs/GCE	Enrofloxacin	Square wave anodic stripping voltammetry	0.041 µmol/L	0.05−10 µmol/L, 10−120 µmol/L	[92]

Table 1. *Cont.*

Electrode	Target Substance	Detection Methods	LOD	Linear Range	References
COF@NH$_2$-CNT/GCE	Furazolidone	Differential pulse voltammetry	7.75×10^{-8} M	0.2–100 µM	[67]
PtNP/COF$_{TFPB\text{-}DHzDS}$@rGO/ePAD	Furazolidone	Differential pulse voltammetry	0.23 µM	0.69–110 µM	[93]
MIP/MoS$_2$/NH$_2$-MWCNT@COF/GCE	Sulfamerazine	Differential pulse voltammetry	1.1×10^{-7} M	3.0×10^{-7}–2.0×10^{-4} M	[94]
Fe$_3$O$_4$@COFs@MIPs/SPE	Tetracycline	Differential pulse voltammetry	2.4×10^{-1} g/mL	1×10^{-10}–1×10^{-4} g/mL	[95]
Co$_3$O$_4$@TAPBDMTP-COF	TBHQ	Differential pulse voltammetry	0.02 µM	0.05–1, 1–4 $\times 10^2$ µM	[97]
COF/AgNPs/CC	Bisphenol A Bisphenol S	Differential pulse voltammetry	0.15 µmol/L 0.15 µmol/L	0.5–100 µmol/L 0.5–100 µmol/L	[99]

4. Conclusions

The purpose of this review has been to highlight the outstanding potential of COFs as emerging porous materials for electrochemical sensing, in particular to ensure food safety. It is becoming increasingly evident that COFs (covalent organic frameworks) are emerging porous materials with a high crystallinity, a good degree of stability, and a controllable pore size and topology. In the field of electrochemical sensing, COFs have exhibited great potential because of their flexible design. Analytes can undergo adsorption and electrochemical reactions facilitated by COFs due to their high surface area and porosity. It is possible to enhance the selectivity and sensitivity of COFs by introducing functional groups. Furthermore, the solid-state nature of COFs makes them easy to integrate with sensing platforms, enhancing their repeatability and stability.

Food safety can be improved through the use of electrochemical sensors, as they offer a high level of accuracy, simplicity, cost-effectiveness, and rapid response. COF-based electrochemical sensors are providing a platform for ensuring food safety and quality. Various designs and types of COFs have been reported, including B-O structures, imine bonds, C=C structures, triazine structures, and other types of connection. Further, the reports discuss methods for improving the performance of COF-based electrochemical sensors and discuss their application for the detection of food pollutants such as pesticides, heavy metal ions, antibiotics, etc.

In spite of the excellent performance and promising prospects of COF-based electrochemical sensors, they exhibit a few challenges, pertaining to enhanced stability and repeatability, miniaturization, and on-site operation. In spite of these challenges, COF-based electrochemical sensors continue to hold great promise in the field of food analysis. It is anticipated that future research will emphasize the combination of novel COFs and advanced electrochemical technologies in order to provide electrochemical sensors with excellent analytical performance.

As a part of the design of COFs, the appropriate pore size and the specific functional groups will be considered based on the physical and chemical properties of the target analytes, which will enhance the selectivity of the modified electrode. It has been proposed that post-modification methods could be employed in order to modify the prepared COF materials and achieve improved properties. Cui et al. modified the COFs with cyclodextrin in order to enhance their ability to transport amino acids selectively [100]. It was found in the study of Li et al. that carboxylic acid groups added to COF materials significantly enhanced the adsorption of heavy metals such as cadmium and mercury [101]. Consequently, the modification of COF materials will enhance their application in the electrochemical detection industry in the future. Moreover, the post-synthetic modifications and alternative methods of enhancing the stability of COFs, particularly imine-based COFs, will increase the requirement for recycling. Additionally, developing mild and effective methods for the modification of COFs on electrodes is crucial in order to ensure optimal contact and minimize any adverse effects on the efficiency of the detection process.

Electrochemical sensors based on COFs are typically constructed by coating or modifying conventional electrodes, but the direct integration of COFs into the device structure can further enhance their sensing capabilities. The advantage of this approach is that it minimizes the signal-to-noise ratio and increases the surface area for the optimum binding and reaction of analytes. It is also advantageous to integrate multiple COF-based sensors on a single substrate, which allows for multiplexed detection, thus increasing the overall efficiency of the system.

COF-based electrodes have recently been shown to have potential applications in the analysis of biomedical and environmental pollutants. It has been demonstrated, for example, that COF-modified electrodes exhibit superior sensitivity, selectivity, stability, and reproducibility when compared with classical electrodes. In particular, COF-modified electrodes are remarkably recyclable even after multiple cycles, demonstrating their durability. While COFs have made significant advancements in electroanalytical chemistry, their application is still relatively new, and challenges must be overcome in order to develop COF-based electrochemical sensors in the future.

Looking ahead, it is vital to develop the reasonable design and controllable synthesis of multifunctional electrode materials based on COFs in order to improve their selectivity towards food pollutants and to enhance their stability and repeatability while achieving the miniaturization of the sensors. Moreover, the biocompatibility study of COF-based electrochemical sensors should be strengthened in order to expand their applications in the detection of biological food contaminants. Additionally, it would be beneficial to address the issue of electrode surface fouling and extend the lifetime of sensors by designing and synthesizing COF electrode materials with anti-fouling properties.

Electrochemical sensors for COFs will be developed in part by the advancement of synthetic techniques and the development of analytical chemistry, among other factors. With the development of new synthetic techniques, COFs with more stable and controllable structures can be synthesized, enabling further investigation into the potential applications of COFs in areas such as food safety and biology. Analytical chemistry advancements have contributed to improving the performance and functionality of sensors. Since the development of synthetic techniques for COFs, researchers have been able to synthesize COFs with more stable and controlled structures. Recently, a number of new synthetic strategies and methods have been developed, including solvent evaporation, cosolvent synthesis, and interfacial synthesis. Electrochemical sensors for COFs can now be developed using these new techniques, providing more options and possibilities. Electrochemical sensors for COFs must also take into account a number of analytical chemistry considerations. It may be necessary, for example, to select the correct sensor materials and synthesis methods in order to achieve a higher level of sensitivity and selectivity. It is also important to pay attention to performance indicators, including stability, repeatability, and practicality. A further enhancement of the performance and functionality of the sensors can be achieved through the interaction of COFs with other functionalized materials. The development of electrochemical sensors with COFs in the future will be driven by these trends, which will lead to wider applications and an increased market potential.

In summary, COF-based electrochemical sensors are still in the early stages of application in the field of food analysis, but this innovative research field holds great promise. It has been possible to detect antimicrobials, pesticides, and heavy metals using some of these techniques. It is anticipated that continued research into the design and synthesis of COFs, in conjunction with advanced electrochemical technologies, will contribute to the enhancement of the analytical performance of electrochemical sensors and the development of food safety testing systems.

Author Contributions: Writing—original draft: H.Z., M.L. (Minjie Li) and C.C. (Cuilin Cheng); Writing—review and editing: Y.H., S.F. and R.L.; Visualization: G.C.; Investigation: M.L. (Miaomiao Liu) and C.C. (Can Cui); Project administration: J.L. and X.Y.; Funding acquisition: J.L.; Supervision: X.Y., H.Z., M.L. (Minjie Li) and C.C. (Cuilin Cheng). X.Y., H.Z., M.L. (Minjie Li) and C.C. (Cuilin Cheng) contributed equally to this study and should be considered co-first authors. All authors have read and agreed to the published version of the manuscript.

Funding: This study was financially supported by the National Natural Science Foundation of China (No. 31972040), the Science and Technology Plan Program of Beijing (Z221100007122006), and Project assignment of Chongqing Natural Science Foundation (CSTB2023NSCQ-MSX0440).

Data Availability Statement: The data used to support the findings of this study is contained within the article.

Conflicts of Interest: Author Minjie Li was employed by the company Internal Trade Food Science Research Institute Co., Ltd and Nutrition & Health Research Institute, COFCO Corporation, which had the role of writing-original draft preparation in this study. Author Gaofeng Cao were employed by the company COFCO Corporation, which had the role of visualization in this study. Author Miaomiao Liu were employed by the company COFCO Corporation, which had the role of investigation in this study. Author Can Cui employed by the company COFCO Corporation, which had the role of investigation in this study. Author Jia Liu was employed by the company Internal Trade Food Science Research Institute Co., Ltd and Nutrition & Health Research Institute, COFCO Corporation, which had the role of project administration and funding acquisition in this study. The remaining authors declare that the research was conducted in the absence of any commercial or financial relationships that could be construed as a potential conflict of interest.

References

1. Singh, B.K.; Tiwari, S.; Dubey, N.K. Essential oils and their nanoformulations as green preservatives to boost food safety against mycotoxin contamination of food commodities: A review. *J. Sci. Food Agric.* **2021**, *101*, 4879–4890. [CrossRef] [PubMed]
2. Chen, J.; Yu, X.; Qiu, L.; Deng, M.; Dong, R. Study on Vulnerability and Coordination of Water-Energy-Food System in Northwest China. *Sustainability* **2018**, *10*, 3712. [CrossRef]
3. Ma, T.; Wang, H.; Wei, M.; Lan, T.; Wang, J.; Bao, S.; Ge, Q.; Fang, Y.; Sun, X. Application of smart-phone use in rapid food detection, food traceability systems, and personalized diet guidance, making our diet more health. *Food Res. Int.* **2022**, *152*, 110918. [CrossRef] [PubMed]
4. Zhao, F.; He, J.; Li, X.; Bai, Y.; Ying, Y.; Ping, J. Smart plant-wearable biosensor for in-situ pesticide analysis. *Biosens. Bioelectron.* **2020**, *170*, 112636. [CrossRef] [PubMed]
5. Sarker, A.; Kim, J.E.; Islam, A.; Bilal, M.; Rakib, M.R.J.; Nandi, R.; Rahman, M.M.; Islam, T. Heavy metals contamination and associated health risks in food webs-a review focuses on food safety and environmental sustainability in Bangladesh. *Environ. Sci. Pollut. Res. Int.* **2022**, *29*, 3230–3245. [CrossRef] [PubMed]
6. Su, Z.; Li, T.; Wu, D.; Wu, Y.; Li, G. Recent Progress on Single-Molecule Detection Technologies for Food Safety. *J. Agric. Food Chem.* **2022**, *70*, 458–469. [CrossRef]
7. Filho, W.L.; Setti, A.F.F.; Azeiteiro, U.M.; Lokupitiya, E.; Donkor, F.K.; Etim, N.N.; Matandirotya, N.; Olooto, F.M.; Sharifi, A.; Nagy, G.J.; et al. An overview of the interactions between food production and climate change. *Sci. Total Environ.* **2022**, *838*, 156438. [CrossRef] [PubMed]
8. Khan, M.R.; Alammari, A.M.; Aqel, A.; Azam, M. Trace analysis of environmental endocrine disrupting contaminant bisphenol A in canned, glass and polyethylene terephthalate plastic carbonated beverages of diverse flavors and origin. *Food Sci. Technol.* **2021**, *41*, 210–217. [CrossRef]
9. Xiang, Q.; Huangfu, L.; Dong, S.; Ma, Y.; Li, K.; Niu, L.; Bai, Y. Feasibility of atmospheric cold plasma for the elimination of food hazards: Recent advances and future trends. *Crit. Rev. Food Sci. Nutr.* **2023**, *63*, 4431–4449. [CrossRef]
10. Snyder, F. No country is an island in regulating food safety: How the WTO monitors Chinese food safety laws through the Trade Policy Review Mechanism (TPRM). *J. Integr. Agric.* **2015**, *14*, 2142–2156. [CrossRef]
11. Walorczyk, S.; Kopeć, I.; Szpyrka, E. Pesticide Residue Determination by Gas Chromatography-Tandem Mass Spectrometry as Applied to Food Safety Assessment on the Example of Some Fruiting Vegetables. *Food Anal. Methods* **2015**, *9*, 1155–1172. [CrossRef]
12. Nicolich, R.S.; Werneck-Barroso, E.; Marques, M.A.S. Food safety evaluation: Detection and confirmation of chloramphenicol in milk by high performance liquid chromatography-tandem mass spectrometry. *Anal. Chim. Acta* **2006**, *565*, 97–102. [CrossRef]
13. Pico, Y.; Font, G.; Ruiz, M.J.; Fernandez, M. Control of pesticide residues by liquid chromatography-mass spectrometry to ensure food safety. *Mass. Spectrom. Rev.* **2006**, *25*, 917–960. [CrossRef] [PubMed]
14. Hingmire, S.; Oulkar, D.P.; Utture, S.C.; Ahammed Shabeer, T.P.; Banerjee, K. Residue analysis of fipronil and difenoconazole in okra by liquid chromatography tandem mass spectrometry and their food safety evaluation. *Food Chem.* **2015**, *176*, 145–151. [CrossRef] [PubMed]
15. Jara, M.D.L.; Alvarez, L.A.C.; Guimaraes, M.C.C.; Antunes, P.W.P.; de Oliveira, J.P. Lateral flow assay applied to pesticides detection: Recent trends and progress. *Environ. Sci. Pollut. Res. Int.* **2022**, *29*, 46487–46508. [CrossRef] [PubMed]
16. Du, H.; Xie, Y.; Wang, J. Nanomaterial-sensors for herbicides detection using electrochemical techniques and prospect applications. *TrAC Trends Anal. Chem.* **2021**, *135*, 116178. [CrossRef]
17. Xue, R.; Kang, T.-F.; Lu, L.-P.; Cheng, S.-Y. Electrochemical Sensor Based on the Graphene-Nafion Matrix for Sensitive Determination of Organophosphorus Pesticides. *Anal. Lett.* **2013**, *46*, 131–141. [CrossRef]

18. Zheng, Y.; Mao, S.; Zhu, J.; Fu, L.; Moghadam, M. A scientometric study on application of electrochemical sensors for detection of pesticide using graphene-based electrode modifiers. *Chemosphere* **2022**, *307*, 136069. [CrossRef]
19. Singh, V.V. Recent Advances in Electrochemical Sensors for Detecting Weapons of Mass Destruction. A Review. *Electroanalysis* **2016**, *28*, 920–935. [CrossRef]
20. Xue, R.; Liu, Y.S.; Huang, S.L.; Yang, G.Y. Recent Progress of Covalent Organic Frameworks Applied in Electrochemical Sensors. *ACS Sens.* **2023**, *8*, 2124–2148. [CrossRef]
21. Fukatsu, A.; Kondo, M.; Masaoka, S. Electrochemical measurements of molecular compounds in homogeneous solution under photoirradiation. *Coord. Chem. Rev.* **2018**, *374*, 416–429. [CrossRef]
22. Ma, X.; Pang, C.; Li, S.; Xiong, Y.; Li, J.; Luo, J.; Yang, Y. Synthesis of Zr-coordinated amide porphyrin-based two-dimensional covalent organic framework at liquid-liquid interface for electrochemical sensing of tetracycline. *Biosens. Bioelectron.* **2019**, *146*, 111734. [CrossRef]
23. Wang, M.; Hu, M.; Liu, J.; Guo, C.; Peng, D.; Jia, Q.; He, L.; Zhang, Z.; Du, M. Covalent organic framework-based electrochemical aptasensors for the ultrasensitive detection of antibiotics. *Biosens. Bioelectron.* **2019**, *132*, 8–16. [CrossRef]
24. Yan, X.; Song, Y.; Liu, J.; Zhou, N.; Zhang, C.; He, L.; Zhang, Z.; Liu, Z. Two-dimensional porphyrin-based covalent organic framework: A novel platform for sensitive epidermal growth factor receptor and living cancer cell detection. *Biosens. Bioelectron.* **2019**, *126*, 734–742. [CrossRef]
25. Shu, Y.; Su, T.; Lu, Q.; Shang, Z.; Xu, Q.; Hu, X. Highly Stretchable Wearable Electrochemical Sensor Based on Ni-Co MOF Nanosheet-Decorated Ag/rGO/PU Fiber for Continuous Sweat Glucose Detection. *Anal. Chem.* **2021**, *93*, 16222–16230. [CrossRef] [PubMed]
26. Ling, P.; Lei, J.; Zhang, L.; Ju, H. Porphyrin-encapsulated metal-organic frameworks as mimetic catalysts for electrochemical DNA sensing via allosteric switch of hairpin DNA. *Anal. Chem.* **2015**, *87*, 3957–3963. [CrossRef] [PubMed]
27. Ramajayam, K.; Ganesan, S.; Ramesh, P.; Beena, M.; Kokulnathan, T.; Palaniappan, A. Molecularly Imprinted Polymer-Based Biomimetic Systems for Sensing Environmental Contaminants, Biomarkers, and Bioimaging Applications. *Biomimetics* **2023**, *8*, 245. [CrossRef] [PubMed]
28. Zhou, B.; Sheng, X.; Xie, H.; Zhou, S.; Huang, L.; Zhang, Z.; Zhu, Y.; Zhong, M. Molecularly Imprinted Electrochemistry Sensor Based on AuNPs/RGO Modification for Highly Sensitive and Selective Detection of Nitrofurazone. *Food Anal. Methods* **2023**, *16*, 709–720. [CrossRef]
29. Svalova, T.S.; Saigushkina, A.A.; Verbitskiy, E.V.; Chistyakov, K.A.; Varaksin, M.V.; Rusinov, G.L.; Charushin, V.N.; Kozitsina, A.N. Rapid and sensitive determination of nitrobenzene in solutions and commercial honey samples using a screen-printed electrode modified by 1,3-/1,4-diazines. *Food Chem.* **2022**, *372*, 131279. [CrossRef] [PubMed]
30. Van Dersarl, J.J.; Mercanzini, A.; Renaud, P. Integration of 2D and 3D Thin Film Glassy Carbon Electrode Arrays for Electrochemical Dopamine Sensing in Flexible Neuroelectronic Implants. *Adv. Funct. Mater.* **2014**, *25*, 78–84. [CrossRef]
31. Wu, T.; Alharbi, A.; Kiani, R.; Shahrjerdi, D. Quantitative Principles for Precise Engineering of Sensitivity in Graphene Electrochemical Sensors. *Adv. Mater.* **2018**, *31*, 1805752. [CrossRef] [PubMed]
32. Zhang, M.; Guo, W. Simultaneous electrochemical detection of multiple heavy metal ions in milk based on silica-modified magnetic nanoparticles. *Food Chem.* **2023**, *406*, 135034. [CrossRef] [PubMed]
33. Cesewski, E.; Johnson, B.N. Electrochemical biosensors for pathogen detection. *Biosens. Bioelectron.* **2020**, *159*, 112214. [CrossRef] [PubMed]
34. Jain, U.; Saxena, K.; Hooda, V.; Balayan, S.; Singh, A.P.; Tikadar, M.; Chauhan, N. Emerging vistas on pesticides detection based on electrochemical biosensors—An update. *Food Chem.* **2022**, *371*, 131126. [CrossRef]
35. Yue, X.; Luo, X.; Zhou, Z.; Bai, Y. Selective electrochemical determination of tertiary butylhydroquinone in edible oils based on an in-situ assembly molecularly imprinted polymer sensor. *Food Chem.* **2019**, *289*, 84–94. [CrossRef]
36. Yu, L.; Sun, L.; Zhang, Q.; Zhou, Y.; Zhang, J.; Yang, B.; Xu, B.; Xu, Q. Nanomaterials-Based Ion-Imprinted Electrochemical Sensors for Heavy Metal Ions Detection: A Review. *Biosensors* **2022**, *12*, 1096. [CrossRef]
37. Nile, S.H.; Baskar, V.; Selvaraj, D.; Nile, A.; Xiao, J.; Kai, G. Nanotechnologies in Food Science: Applications, Recent Trends, and Future Perspectives. *Nanomicro Lett.* **2020**, *12*, 45. [CrossRef]
38. Park, J.; Hwang, J.C.; Kim, G.G.; Park, J.U. Flexible electronics based on one-dimensional and two-dimensional hybrid nanomaterials. *InfoMat* **2019**, *2*, 33–56. [CrossRef]
39. Sebastian, V.; Arruebo, M.; Santamaria, J. Reaction engineering strategies for the production of inorganic nanomaterials. *Small* **2014**, *10*, 835–853. [CrossRef]
40. Liu, T.; Chu, Z.; Jin, W. Electrochemical mercury biosensors based on advanced nanomaterials. *J. Mater. Chem. B* **2019**, *7*, 3620–3632. [CrossRef]
41. Wei, Q.; Xiong, F.; Tan, S.; Huang, L.; Lan, E.H.; Dunn, B.; Mai, L. Porous One-Dimensional Nanomaterials: Design, Fabrication and Applications in Electrochemical Energy Storage. *Adv. Mater.* **2017**, *29*, 1602300. [CrossRef]
42. Ma, Y.; Li, B.; Yang, S. Ultrathin two-dimensional metallic nanomaterials. *Mater. Chem. Front.* **2018**, *2*, 456–467. [CrossRef]
43. Liu, Y.; Pang, H.; Wang, X.; Yu, S.; Chen, Z.; Zhang, P.; Chen, L.; Song, G.; Saleh Alharbi, N.; Omar Rabah, S.; et al. Zeolitic imidazolate framework-based nanomaterials for the capture of heavy metal ions and radionuclides: A review. *Chem. Eng. J.* **2021**, *406*, 127139. [CrossRef]

44. Wang, S.; McGuirk, C.M.; d'Aquino, A.; Mason, J.A.; Mirkin, C.A. Metal-Organic Framework Nanoparticles. *Adv. Mater.* **2018**, *30*, e1800202. [CrossRef] [PubMed]

45. Zhang, X.; Hou, L.; Samori, P. Coupling carbon nanomaterials with photochromic molecules for the generation of optically responsive materials. *Nat. Commun.* **2016**, *7*, 11118. [CrossRef] [PubMed]

46. Geng, K.; He, T.; Liu, R.; Dalapati, S.; Tan, K.T.; Li, Z.; Tao, S.; Gong, Y.; Jiang, Q.; Jiang, D. Covalent Organic Frameworks: Design, Synthesis, and Functions. *Chem. Rev.* **2020**, *120*, 8814–8933. [CrossRef]

47. Cote, A.P.; Benin, A.I.; Ockwig, N.W.; O'Keeffe, M.; Matzger, A.J.; Yaghi, O.M. Porous, Crystalline, Covalent Organic Frameworks. *Science* **2005**, *310*, 1166–1170. [CrossRef]

48. El-Kaderi, H.M.; Hunt, J.R.; Mendoza-Cortés, J.L.; Côté, A.P.; Taylor, R.E.; O'Keeffe, M.; Yaghi, O.M. Designed Synthesis of 3D Covalent Organic Frameworks. *Science* **2007**, *316*, 268–272. [CrossRef] [PubMed]

49. Ding, S.Y.; Wang, W. Covalent organic frameworks (COFs): From design to applications. *Chem. Soc. Rev.* **2013**, *42*, 548–568. [CrossRef] [PubMed]

50. Hu, F.; Xue, Y.; Xu, J.; Lu, B. PEDOT-Based Conducting Polymer Actuators. *Front. Robot. AI* **2019**, *6*, 114. [CrossRef] [PubMed]

51. Díaz, U.; Corma, A. Ordered covalent organic frameworks, COFs and PAFs. From preparation to application. *Coord. Chem. Rev.* **2016**, *311*, 85–124. [CrossRef]

52. Xu, F.; Xu, H.; Chen, X.; Wu, D.; Wu, Y.; Liu, H.; Gu, C.; Fu, R.; Jiang, D. Radical covalent organic frameworks: A general strategy to immobilize open-accessible polyradicals for high-performance capacitive energy storage. *Angew. Chem. Int. Ed. Engl.* **2015**, *54*, 6814–6818. [CrossRef]

53. Chen, X.; Addicoat, M.; Irle, S.; Nagai, A.; Jiang, D. Control of crystallinity and porosity of covalent organic frameworks by managing interlayer interactions based on self-complementary pi-electronic force. *J. Am. Chem. Soc.* **2013**, *135*, 546–549. [CrossRef] [PubMed]

54. Xiang, Z.; Cao, D.; Huang, L.; Shui, J.; Wang, M.; Dai, L. Nitrogen-Doped Holey Graphitic Carbon from 2D Covalent Organic Polymers for Oxygen Reduction. *Adv. Mater.* **2014**, *26*, 3315–3320. [CrossRef] [PubMed]

55. Lanni, L.M.; Tilford, R.W.; Bharathy, M.; Lavigne, J.J. Enhanced hydrolytic stability of self-assembling alkylated two-dimensional covalent organic frameworks. *J. Am. Chem. Soc.* **2011**, *133*, 13975–13983. [CrossRef] [PubMed]

56. Vyas, V.S.; Vishwakarma, M.; Moudrakovski, I.; Haase, F.; Savasci, G.; Ochsenfeld, C.; Spatz, J.P.; Lotsch, B.V. Exploiting Noncovalent Interactions in an Imine-Based Covalent Organic Framework for Quercetin Delivery. *Adv. Mater.* **2016**, *28*, 8749–8754. [CrossRef] [PubMed]

57. Uribe-Romo, F.J.; Hunt, J.R.; Furukawa, H.; Klock, C.; O'Keeffe, M.; Yaghi, O.M. A Crystalline Imine-Linked 3-D Porous Covalent Organic Framework. *J. Am. Chem. Soc.* **2009**, *131*, 4570–4571. [CrossRef]

58. Uribe-Romo, F.J.; Doonan, C.J.; Furukawa, H.; Oisaki, K.; Yaghi, O.M. Crystalline covalent organic frameworks with hydrazone linkages. *J. Am. Chem. Soc.* **2011**, *133*, 11478–11481. [CrossRef]

59. Kandambeth, S.; Shinde, D.B.; Panda, M.K.; Lukose, B.; Heine, T.; Banerjee, R. Enhancement of chemical stability and crystallinity in porphyrin-containing covalent organic frameworks by intramolecular hydrogen bonds. *Angew. Chem. Int. Ed. Engl.* **2013**, *52*, 13052–13056. [CrossRef]

60. Zhuang, X.; Zhao, W.; Zhang, F.; Cao, Y.; Liu, F.; Bi, S.; Feng, X. A two-dimensional conjugated polymer framework with fully sp2-bonded carbon skeleton. *Polym. Chem.* **2016**, *7*, 4176–4181. [CrossRef]

61. Kuhn, P.; Thomas, A.; Antonietti, M. Toward Tailorable Porous Organic Polymer Networks: A High-Temperature Dynamic Polymerization Scheme Based on Aromatic Nitriles. *Macromolecules* **2009**, *42*, 319–326. [CrossRef]

62. Grill, L.; Dyer, M.; Lafferentz, L.; Persson, M.; Peters, M.V.; Hecht, S. Nano-architectures by covalent assembly of molecular building blocks. *Nat. Nanotechnol.* **2007**, *2*, 687–691. [CrossRef]

63. Jiang, S.Y.; Gan, S.X.; Zhang, X.; Li, H.; Qi, Q.Y.; Cui, F.Z.; Lu, J.; Zhao, X. Aminal-Linked Covalent Organic Frameworks through Condensation of Secondary Amine with Aldehyde. *J. Am. Chem. Soc.* **2019**, *141*, 14981–14986. [CrossRef] [PubMed]

64. Fang, Q.; Zhuang, Z.; Gu, S.; Kaspar, R.B.; Zheng, J.; Wang, J.; Qiu, S.; Yan, Y. Designed synthesis of large-pore crystalline polyimide covalent organic frameworks. *Nat. Commun.* **2014**, *5*, 4503. [CrossRef] [PubMed]

65. Zhao, C.; Lyu, H.; Ji, Z.; Zhu, C.; Yaghi, O.M. Ester-Linked Crystalline Covalent Organic Frameworks. *J. Am. Chem. Soc.* **2020**, *142*, 14450–14454. [CrossRef] [PubMed]

66. Guo, J.; Xu, Y.; Jin, S.; Chen, L.; Kaji, T.; Honsho, Y.; Addicoat, M.A.; Kim, J.; Saeki, A.; Ihee, H.; et al. Conjugated organic framework with three-dimensionally ordered stable structure and delocalized pi clouds. *Nat. Commun.* **2013**, *4*, 2736. [CrossRef] [PubMed]

67. Sun, Y.; Waterhouse, G.I.N.; Xu, L.; Qiao, X.; Xu, Z. Three-dimensional electrochemical sensor with covalent organic framework decorated carbon nanotubes signal amplification for the detection of furazolidone. *Sens. Actuators B Chem.* **2020**, *321*, 128501. [CrossRef]

68. Wang, L.; Xie, Y.; Yang, Y.; Liang, H.; Wang, L.; Song, Y. Electroactive Covalent Organic Frameworks/Carbon Nanotubes Composites for Electrochemical Sensing. *ACS Appl. Nano Mater.* **2020**, *3*, 1412–1419. [CrossRef]

69. Lu, Z.; Shi, Z.; Huang, S.; Zhang, R.; Li, G.; Hu, Y. Covalent organic framework derived Fe(3)O(4)/N co-doped hollow carbon nanospheres modified electrode for simultaneous determination of biomolecules in human serum. *Talanta* **2020**, *214*, 120864. [CrossRef]

70. Xu, M.; Wang, L.; Xie, Y.; Song, Y.; Wang, L. Ratiometric electrochemical sensing and biosensing based on multiple redox-active state COFDHTA-TTA. *Sens. Actuators B Chem.* **2019**, *281*, 1009–1015. [CrossRef]
71. Zhang, T.; Chen, Y.; Huang, W.; Wang, Y.; Hu, X. A novel AuNPs-doped COFs composite as electrochemical probe for chlorogenic acid detection with enhanced sensitivity and stability. *Sens. Actuators B Chem.* **2018**, *276*, 362–369. [CrossRef]
72. Pan, F.; Tong, C.; Wang, Z.; Han, H.; Liu, P.; Pan, D.; Zhu, R. Nanocomposite based on graphene and intercalated covalent organic frameworks with hydrosulphonyl groups for electrochemical determination of heavy metal ions. *Mikrochim. Acta* **2021**, *188*, 295. [CrossRef]
73. Zhang, T.; Ma, N.; Ali, A.; Wei, Q.; Wu, D.; Ren, X. Electrochemical ultrasensitive detection of cardiac troponin I using covalent organic frameworks for signal amplification. *Biosens. Bioelectron.* **2018**, *119*, 176–181. [CrossRef]
74. Zhang, T.; Gao, C.; Huang, W.; Chen, Y.; Wang, Y.; Wang, J. Covalent organic framework as a novel electrochemical platform for highly sensitive and stable detection of lead. *Talanta* **2018**, *188*, 578–583. [CrossRef]
75. Wang, J.; Qu, X.; Zhao, L.; Yan, B. Fabricating Nanosheets and Ratiometric Detection of 5-Fluorouracil by Covalent Organic Framework Hybrid Material. *Anal. Chem.* **2021**, *93*, 4308–4316. [CrossRef] [PubMed]
76. Pundir, C.S.; Malik, A. Bio-sensing of organophosphorus pesticides: A review. *Biosens. Bioelectron.* **2019**, *140*, 111348. [CrossRef] [PubMed]
77. Xiao, Y.; Wu, N.; Wang, L.; Chen, L. A Novel Paper-Based Electrochemical Biosensor Based on N,O-Rich Covalent Organic Frameworks for Carbaryl Detection. *Biosensors* **2022**, *12*, 899. [CrossRef] [PubMed]
78. Wang, L.; Wu, N.; Wang, L.; Song, Y.; Ma, G. Accurate detection of organophosphorus pesticides based on covalent organic framework nanofiber with a turn-on strategy. *Sens. Actuators B Chem.* **2022**, *372*, 132608. [CrossRef]
79. Niu, K.; Zhang, Y.; Chen, J.; Lu, X. 2D Conductive Covalent Organic Frameworks with Abundant Carbonyl Groups for Electrochemical Sensing. *ACS Sens.* **2022**, *7*, 3551–3559. [CrossRef] [PubMed]
80. Liu, Y.; Zhou, M.; Jin, C.; Zeng, J.; Huang, C.; Song, Q.; Song, Y. Preparation of a Sensor Based on Biomass Porous Carbon/Covalent-Organic Frame Composites for Pesticide Residues Detection. *Front. Chem.* **2020**, *8*, 643. [CrossRef]
81. Wang, X.; Yang, S.; Shan, J.; Bai, X. Novel Electrochemical Acetylcholinesterase Biosensor Based on Core-Shell Covalent Organic Framework@Multi-Walled Carbon Nanotubes (COF@MWCNTs) Composite for Detection of Malathion. *Int. J. Electrochem. Sci.* **2022**, *17*, 220543. [CrossRef]
82. Jin, J.; Zhao, X.; Zhang, L.; Hu, Y.; Zhao, J.; Tian, J.; Ren, J.; Lin, K.; Cui, C. Heavy metals in daily meals and food ingredients in the Yangtze River Delta and their probabilistic health risk assessment. *Sci. Total Environ.* **2023**, *854*, 158713. [CrossRef]
83. Xiong, B.; Xu, T.; Li, R.; Johnson, D.; Ren, D.; Liu, H.; Xi, Y.; Huang, Y. Heavy metal accumulation and health risk assessment of crayfish collected from cultivated and uncultivated ponds in the Middle Reach of Yangtze River. *Sci. Total Environ.* **2020**, *739*, 139963. [CrossRef]
84. Ugochukwu, U.C.; Chukwuone, N.; Jidere, C.; Ezeudu, B.; Ikpo, C.; Ozor, J. Heavy metal contamination of soil, sediment and water due to galena mining in Ebonyi State Nigeria: Economic costs of pollution based on exposure health risks. *J. Environ. Manag.* **2022**, *321*, 115864. [CrossRef]
85. Zhang, H.; Gao, Z.; Liu, Y.; Ran, C.; Mao, X.; Kang, Q.; Ao, W.; Fu, J.; Li, J.; Liu, G.; et al. Microwave-assisted pyrolysis of textile dyeing sludge, and migration and distribution of heavy metals. *J. Hazard. Mater.* **2018**, *355*, 128–135. [CrossRef]
86. Pei, L.; Su, J.; Yang, H.; Wu, Y.; Du, Y.; Zhu, Y. A novel covalent-organic framework for highly sensitive detection of Cd^{2+}, Pb^{2+}, Cu^{2+} and Hg^{2+}. *Microporous Mesoporous Mater.* **2022**, *333*, 111742. [CrossRef]
87. Han, J.; Pei, L.; Du, Y.; Zhu, Y. Tripolycyanamide-2,4,6-triformyl pyrogallol covalent organic frameworks with many coordination sites for detection and removal of heavy metal ions. *J. Ind. Eng. Chem.* **2022**, *107*, 53–60. [CrossRef]
88. Jalali Sarvestani, M.R.; Madrakian, T.; Afkhami, A. Simultaneous determination of Pb(2+) and Hg(2+) at food specimens by a Melamine-based covalent organic framework modified glassy carbon electrode. *Food Chem.* **2023**, *402*, 134246. [CrossRef] [PubMed]
89. Jalali Sarvestani, M.R.; Madrakian, T.; Afkhami, A. Ultra-trace levels voltammetric determination of Pb(2+) in the presence of Bi(3+) at food samples by a Fe(3)O(4)@Schiff base Network(1) modified glassy carbon electrode. *Talanta* **2022**, *250*, 123716. [CrossRef] [PubMed]
90. Fernandes, P.; Martens, E. Antibiotics in late clinical development. *Biochem. Pharmacol.* **2017**, *133*, 152–163. [CrossRef] [PubMed]
91. Magalhaes, D.; Freitas, A.; Sofia Vila Pouca, A.; Barbosa, J.; Ramos, F. The use of ultra-high-pressure-liquid-chromatography tandem time-of-flight mass spectrometry as a confirmatory method in drug residue analysis: Application to the determination of antibiotics in piglet liver. *J. Chromatogr. B Anal. Technol. Biomed. Life Sci.* **2020**, *1153*, 122264. [CrossRef]
92. Lu, S.; Wang, S.; Wu, P.; Wang, D.; Yi, J.; Li, L.; Ding, P.; Pan, H. A composite prepared from covalent organic framework and gold nanoparticles for the electrochemical determination of enrofloxacin. *Adv. Powder Technol.* **2021**, *32*, 2106–2115. [CrossRef]
93. Chen, R.; Peng, X.; Song, Y.; Du, Y. A Paper-Based Electrochemical Sensor Based on PtNP/COF(TFPB-DHzDS)@rGO for Sensitive Detection of Furazolidone. *Biosensors* **2022**, *12*, 904. [CrossRef] [PubMed]
94. Sun, Y.; Xu, L.; Waterhouse, G.I.N.; Wang, M.; Qiao, X.; Xu, Z. Novel three-dimensional electrochemical sensor with dual signal amplification based on MoS2 nanosheets and high-conductive NH2-MWCNT@COF for sulfamerazine determination. *Sens. Actuators B Chem.* **2019**, *281*, 107–114. [CrossRef]

95. Yang, Y.; Shi, Z.; Wang, X.; Bai, B.; Qin, S.; Li, J.; Jing, X.; Tian, Y.; Fang, G. Portable and on-site electrochemical sensor based on surface molecularly imprinted magnetic covalent organic framework for the rapid detection of tetracycline in food. *Food Chem.* **2022**, *395*, 133532. [CrossRef]

96. Kashanian, S.; Dolatabadi, J.E.N. DNA binding studies of 2-tert-butylhydroquinone (TBHQ) food additive. *Food Chem.* **2009**, *116*, 743–747. [CrossRef]

97. Chen, Y.; Xie, Y.; Sun, X.; Wang, Y.; Wang, Y. Tunable construction of crystalline and shape-tailored Co$_3$O$_4$@TAPB-DMTP-COF composites for the enhancement of tert-butylhydroquinone electrocatalysis. *Sens. Actuators B Chem.* **2021**, *331*, 129438. [CrossRef]

98. Deceuninck, Y.; Bichon, E.; Geny, T.; Veyrand, B.; Grandin, F.; Viguie, C.; Marchand, P.; Le Bizec, B. Quantitative method for conjugated metabolites of bisphenol A and bisphenol S determination in food of animal origin by Ultra High Performance Liquid Chromatography-Tandem Mass Spectrometry. *J. Chromatogr. A* **2019**, *1601*, 232–242. [CrossRef]

99. Pang, Y.H.; Wang, Y.Y.; Shen, X.F.; Qiao, J.Y. Covalent organic framework modified carbon cloth for ratiometric electrochemical sensing of bisphenol A and S. *Mikrochim. Acta* **2022**, *189*, 189. [CrossRef]

100. Yuan, C.; Wu, X.; Gao, R.; Han, X.; Liu, Y.; Long, Y.; Cui, Y. Nanochannels of Covalent Organic Frameworks for Chiral Selective Transmembrane Transport of Amino Acids. *J. Am. Chem. Soc.* **2019**, *141*, 20187–20197. [CrossRef]

101. Zhu, R.; Zhang, P.; Zhang, X.; Yang, M.; Zhao, R.; Liu, W.; Li, Z. Fabrication of synergistic sites on an oxygen-rich covalent organic framework for efficient removal of Cd(II) and Pb(II) from water. *J. Hazard. Mater.* **2022**, *424*, 127301. [CrossRef] [PubMed]

Review

A Comprehensive Review of Pesticide Residues in Peppers

Jae-Han Shim [1,*], Jong-Bang Eun [2], Ahmed A. Zaky [3], Ahmed S. Hussein [3], Ahmet Hacimüftüoğlu [4,5] and A. M. Abd El-Aty [4,6,*]

1   Natural Products Chemistry Laboratory, Biotechnology Research Institute, Chonnam National University, 77 Yongbong-ro, Buk-gu, Gwangju 500-757, Republic of Korea
2   Department of Food Science and Technology, Chonnam National University, Gwangju 61186, Republic of Korea
3   Department of Food Technology, Food Industries and Nutrition Research Institute, National Research Centre, Cairo 12622, Egypt
4   Department of Medical Pharmacology, Medical Faculty, Ataturk University, Erzurum 25240, Turkey
5   Vaccine Development Application and Research Center, Ataturk University, Erzurum 25240, Turkey
6   Department of Pharmacology, Faculty of Veterinary Medicine, Cairo University, Giza 12211, Egypt
*   Correspondence: jhshim@jnu.ac.kr (J.-H.S.); abdelaty44@hotmail.com (A.M.A.)

Citation: Shim, J.-H.; Eun, J.-B.; Zaky, A.A.; Hussein, A.S.; Hacimüftüoğlu, A.; Abd El-Aty, A.M. A Comprehensive Review of Pesticide Residues in Peppers. *Foods* **2023**, *12*, 970. https://doi.org/10.3390/foods12050970

Academic Editor: Amin Mousavi Khaneghah

Received: 22 January 2023
Revised: 14 February 2023
Accepted: 21 February 2023
Published: 24 February 2023

Abstract: Pesticides are chemicals that are used to control pests such as insects, fungi, and weeds. Pesticide residues can remain on crops after application. Peppers are popular and versatile foods that are valued for their flavor, nutrition, and medicinal properties. The consumption of raw or fresh peppers (bell and chili) can have important health benefits due to their high levels of vitamins, minerals, and antioxidants. Therefore, it is crucial to consider factors such as pesticide use and preparation methods to fully realize these benefits. Ensuring that the levels of pesticide residues in peppers are not harmful to human health requires rigorous and continuous monitoring. Several analytical methods, such as gas chromatography (GC), liquid chromatography (LC), mass spectrometry (MS), infrared spectroscopy (IR), ultraviolet–visible spectroscopy (UV–Vis), and nuclear magnetic resonance spectroscopy (NMR), can detect and quantify pesticide residues in peppers. The choice of analytical method depends on the specific pesticide, that is being tested for and the type of sample being analyzed. The sample preparation method usually involves several processes. This includes extraction, which is used to separate the pesticides from the pepper matrix, and cleanup, which removes any interfering substances that could affect the accuracy of the analysis. Regulatory agencies or food safety organizations typically monitor pesticide residues in peppers by stipulating maximum residue limits (MRLs). Herein, we discuss various sample preparation, cleanup, and analytical techniques, as well as the dissipation patterns and application of monitoring strategies for analyzing pesticides in peppers to help safeguard against potential human health risks. From the authors' perspective, several challenges and limitations exist in the analytical approach to monitoring pesticide residues in peppers. These include the complexity of the matrix, the limited sensitivity of some analytical methods, cost and time, a lack of standard methods, and limited sample size. Furthermore, developing new analytical methods, using machine learning and artificial intelligence, promoting sustainable and organic growing practices, improving sample preparation methods, and increasing standardization could assist efficiently in analyzing pesticide residues in peppers.

Keywords: pepper; pesticide residue; monitoring; analytical approach; maximum residue limit; dissipation pattern

1. Introduction

Peppers are an important part of many diets worldwide due to their nutritional value. Peppers are a rich source of vitamins and minerals, including vitamins C and A and potassium [1]. They are also high in antioxidants and phytochemicals [2], which are associated with many health benefits, including boosting the immune system [3], maintaining healthy skin and eyesight [4], reducing the risk of chronic diseases, such as cancer and cardiovascular disease [3], and managing blood sugar levels [4]. Furthermore, peppers have a distinctive flavor [5] and can be used in various dishes, including salads, sandwiches, and cooked dishes. They can also add color and flavor to various dishes, including soups, stews, and stir-fries. Moreover, peppers are an important part of the cuisine and culture of many countries worldwide. They are often key ingredients in traditional dishes, such as curries, stews, and sauces. In the Republic of Korea, chili peppers are frequently used in traditional cuisines, such as kimchi [6]. They are available in various forms, including fresh, cooked, dried powder, processed, and frozen, and there are various types of pepper, including sweet peppers and hot peppers. Some common examples include bell peppers, banana peppers, Cubanelle peppers, Jalapeño peppers, Habanero peppers, and Serrano peppers [7,8]. Peppers are also an important source of income for many small farmers, particularly in developing countries where they are grown for local consumption and export and contribute to the economic development of communities. According to the Food and Agriculture Organization of the United Nations (FAO), the worldwide production of pepper was approximately 540,000 metric tons in 2020. The top five pepper-producing countries worldwide are listed in the following order: Vietnam, Indonesia, Brazil, India, and Malaysia [9].

Pepper leaves can be used as an ingredient in a type of Korean dish called *Namul Namul* refers to dishes made with seasoned vegetables and herbs that serve as side dishes or as part of a larger meal [10]. In general, they are not typically consumed as food. However, they can be used in traditional medicine in some cultures. Pepper leaves have been used in traditional remedies to treat a range of ailments, including digestive disorders and respiratory diseases. However, pepper leaves have not been extensively studied for medicinal purposes, and their safety and effectiveness have not yet been established.

The quantity and safety of agricultural products are intimately linked to public health, social stability, and sustainable development. Therefore, this topic has gained increasing concern among the general public, health authorities, and the scientific community [11]. The World Health Organization (WHO) states that fruits and vegetables are crucial to a healthy diet. Reduced consumption of fruits and vegetables could correlate with poor health and a higher risk of noncommunicable diseases.

Many pests, including insects, fungi, and weeds, can affect pepper plants. Insects, such as aphids, whiteflies, and mites, can feed on the leaves and stems of pepper plants, causing damage and reducing yields. Insects can also transmit diseases, such as viruses and bacteria, to pepper plants. Fungal diseases in pepper plants, such as anthracnose and blossom end rot, can cause symptoms such as leaf and fruit discoloration, wilting, and plant tissue death. Additionally, weeds can compete with pepper plants for light, water, and nutrients: reducing the growth and yield of pepper plants. Weeds can also harbor pests and diseases that can affect pepper plants.

Insecticides, herbicides, and fungicides are often used to help increase the yield of pepper crops by reducing the damage caused by pests. This could improve farmers' profitability in terms of pepper production [12]. However, it is important to note that the use of pesticides can also have negative impacts, including risks to human health and the environment [13]. The WHO has reported that approximately three million cases of pesticide poisoning occur annually across the globe, of which 220,000 cases are fatal [14]. As a result of population increase and rapid urbanization, the use of pesticides is also continuously rising [15]. Pesticides can remain on pepper surfaces after their application, and consuming peppers with high pesticide residues can harm human health. In addition, pesticide use can also negatively impact nontarget species, such as birds and bees, and contribute to the decline in these species [16]. It can also lead to the development of resistance in pests, requiring the use of even more pesticides. To mitigate the negative impacts of pesticides, it is vital to use them only when necessary and to follow proper application and safety guidelines. It is also important to monitor and control the levels of pesticide residues in peppers to ensure that they are safe for human consumption.

The intake of raw and cooked vegetables, such as peppers, is one of the most common pesticide exposure routes [17]. The impact of pesticides on the nutritional value of peppers depends on several factors, including the type and amount of pesticide used, the length of time the pesticides remained on the peppers, and their overall nutritional content. In general, pesticides can potentially reduce the nutritional value of peppers by decreasing the levels of certain nutrients, such as vitamins and minerals [18]. Some pesticides may also have toxic effects on beneficial microorganisms in the soil, which could affect the overall nutrient content of peppers.

The current review discusses different extraction and analytical procedures that are used to determine pesticide residues in peppers. Additionally, dissipation patterns and monitoring strategies are also reviewed. A conclusion and potential future perspectives are proposed based on the authors' viewpoints.

2. Maximum Residue Limits

Maximum residue limits (MRLs) are regulatory limits that are set for the levels of pesticide that are allowed to remain on food crops, including peppers, after the application of plant production products and preharvest intervals. MRLs are set to protect human health by ensuring that pesticides on food crops are below the levels that could potentially be harmful. MRLs for pesticides can vary between countries, as each country may have different regulations and guidelines for pesticide use. Some countries may have specific MRLs for particular pesticides in peppers, while others may have more general MRLs for specific pesticides in all food crops. MRLs are typically set based on the results of toxicological assessments, which evaluate the potential health risks of different pesticides [19]. MRLs are typically set at levels well below those that could potentially harm human health. It is important to note that MRLs are not intended to measure pesticide safety.

Disagreements over permissible levels between nations could impede trade globally, thus highlighting the urgent need for MRL standardization. The European Union (EU) and the Codex Alimentarius Commission of the Joint Food and Agricultural Organization of the United Nations (FAO)/WHO [20] have established reference MRLs. Each country uses one of two strategies to limit pesticide residues in agricultural commodities: (a) the regulatory monitoring of agricultural raw materials, measuring the residual levels of particular matrices following the MRL [21,22]; or (b) whole diet research, which analyzes the foods people eat to estimate their dietary intake of pesticides [23–25]. A viable approach is required to identify and measure residues at a level equal to or lower than the MRL (Table 1) and verify the identification of substances in agricultural products for research and regulatory purposes. The fundamental steps in multi-residue methods (MRMs) and single-residue methods (SRMs) are essentially the same. To monitor or screen different kinds of pesticides in specific products, MRMs are typically used. By contrast, SRMs are often used for substances that cannot be determined by MRM methods and require specific procedures for sample preparation and determination [26–29]. Consequently, a suitable analytical technique should be established to quantitatively identify the levels of pesticides in peppers for safety and dietary risk assessment.

Table 1. Pesticide levels of peppers in which the maximum residue limits (MRLs) were exceeded.

Sample Matrix	Pesticide	Efficacy	Peppers MRL		References
			EU	Codex	
Bell Pepper	Bifenthrin	Insecticide, Acaricide	0.5	0.5 (P), 5 (D)	[30]
	Ethoprophos	Nematicide, Insecticide	0.05	0.2 (D)	
	Cypermethrin	Insecticide	0.5	2, (10, D)	
	Cyhalothrin	Insecticide	0.1	0.3 (F), 3 (D)	
	Carbofuran	Insecticide, Nematicide	0.002 *	—	
	Monocrotophos	Insecticide, Acaricide	0.01 *	—	
	Dimethoate	Insecticide, Acaricide	0.01 *	3 (D)	
	Methamidophos	Insecticide, Acaricide	0.01 *	—	
Green Pepper	Acetamiprid	Insecticide	0.3	0.2 (F), 2 (D)	[31]
	Boscalid	Fungicide	3	3 (F), 10 (D)	
	Azoxystrobin	Fungicide	3	3 (F), 30 (D)	
	Triadimenol	Fungicide	0.5	1 (F), (5 D)	
	Cyprodinil	Fungicide	1.5	2 (F), (9, D)	
	Metalaxyl	Fungicide	0.5 (Including Metalaxyl—M)	1 (P), (10 D)	
	Spinosad	Insecticide	2	0.3 P, (3 D)	
	Tebuconazole	Fungicide	0.6	10 (D)	
	Thiamethoxam	Insecticide	0.7	0.7 (F), (7 D)	
Pepper	Hexaconazole	Fungicide	0.01 *	—	[32]
	Propiconazole	Fungicide	0.01 *	—	
Chili pepper	Methomyl	Insecticide, Acaricide	0.04	0.7 (P), (10 D)	[33]
	Imidacloprid	Insecticide	0.9	1 (P), (10 D)	
	Metalaxyl	Fungicide	0.5 (Including Metalaxyl—M)	1 (P), (10 D)	
	Cyproconazole	Fungicide	0.05 *	—	
Greenhouse sweet pepper	Azoxystrobin	Fungicide	3	3 (F), 30 (D)	[34]
	Iprodione	Fungicide	0.01 *	—	
	Pyrimethanil	Fungicide	2	—	
Field sweet Pepper	Boscalid	Fungicide	3	3 (F), 10 (D)	[34]
Pepper	Napropamide	Herbicide	0.01 *	—	[35]
	Pendimethalin	Herbicide	0.05 *	—	
	Trifluralin	Herbicide	0.01 *	—	
	Diazinon	Insecticide, Acaricide	0.05	0.5 (D)	
	Malathion	Insecticide, Acaricide	0.02 *	0.1 (P), (1 D)	
	Pirimiphos—methyl	Insecticide, Acaricide	0.01 *	—	
Green pepper	Chlorpyrifos	Insecticide	0.01 *	20 (D)	[36]
	Chlorfenapyr	Insecticide, Acaricide	0.01 *	0.3 (P), (3 D)	
	Tebuconazole	Fungicide	0.6	10 (D)	
	Methamidophos	Insecticide, Acaricide	0.01 *	—	
	Cypermethrin	Insecticide	0.5	2, (10, D)	
Dried red pepper	Pencycuron	Fungicide	0.02 *	—	[24]

Table 1. *Cont.*

Sample Matrix	Pesticide	Efficacy	Peppers MRL		References
			EU	Codex	
Green pepper	Kresoxim—methyl	Fungicide	0.8	—	
	Diazinon	Insecticide, Acaricide	0.05	0.5 (D)	
	Amitrole	Herbicide	0.01 *	—	[37]
	Bromoxynil	Herbicide	0.01 *	—	
	Carbaryl	Insecticide, plant growth regulator	0.01*	0.5, (2 D)	
	Carbofuran	Insecticide, nematicide	0.002 *	—	
	Chlorpyrifos	Insecticide	0.01 *	20 (D)	
	Dicofol	Acaricide	0.02 *	—	
	Malathion	Insecticide, Acaricide	0.02 *	0.1 (P), (1 D)	
	Metalaxyl	Fungicide	0.5 (Including Metalaxyl—M)	1 (P), (10 D)	
	Methoxychlor	Insecticide	0.01 *	—	
	Paraquat	Herbicide	0.02 *	0.05 (F)	
	Propoxur	Insecticide	0.05 *	—	
	Pyrethrin 1	Insecticide, Acaricide	1	0.05 (P), (0.5 D)	
	Tefluthrin	Insecticide	0.01 *	—	
	Tolclofos—methyl	Fungicide	0.01 *	—	
Green pepper	Endosulfan	Insecticide, Acaricide	0.05 *	—	[38]
	Azinphos-methyl	Insecticide	0.01 *	—	
	Carbaryl	Insecticide, plant growth regulator	0.01 *	0.5, (2 D)	
	Carbofuran	Insecticide, Nematicide	0.002 *	—	
	Chlorothalonil	Fungicide	0.01 *	7 (P), (70, D),	
Green pepper	Dichlorvos	Insecticide, Acaricide	0.01 *	—	[39]
	Omethoate	Insecticide, Acaricide	0.01 *	—	
	Ethoprophos	Nematicide, insecticide	0.05	0.2 (D)	
Red pepper	Chlorpyrifos	Insecticide	0.01 *	20 (D)	[40]
Green pepper, chili pepper	Aldrin	Insecticide	0.01 *	—	[41]
	Alpha—BHC	Insecticide	HCH	—	
	Beta—BHC	Insecticide	HCH	—	
	Chlorothalonil	Fungicide	0.01 *	7 (P), (70, D),	
	Delta—BHC	Insecticide	HCH	—	
	Dieldrin	Insecticide	0.01 *	—	
	Endosulfan	Insecticide, Acaricide	0.05 *	—	
	Heptachlor	Insecticide	0.01 *	—	
	Heptachlor—epoxide	Insecticide	0.01 *	—	
	p,p'—DDT	Insecticide	0.05 *	—	
Pepper	Chlorothalonil	Fungicide	0.01 *	7 (P), (70, D),	[42]
	Endosulfan	Insecticide, Acaricide	0.05 *	—	
	Ethoprophos	Nematicide, insecticide	0.05	0.2 (D)	
	Kresoxim—methyl	Fungicide	0.8	—	

Table 1. *Cont.*

Sample Matrix	Pesticide	Efficacy	Peppers MRL		References
			EU	**Codex**	
Pepper	Pirimiphos—methyl	Insecticide, Acaricide	0.01 *	—	
	Procymidone	Fungicide	0.01 *	—	
	Methyl—chlorpyrifos	Insecticide, Acaricide	0.01 *	20 (D)	[43]
	Ethyl—chlorpyrifos	Insecticide, Acaricide	0.01 *	20 (D)	
	Dimethoate	Insecticide, Acaricide	0.01 *	3 (D)	
Red pepper	Malathion	Insecticide, Acaricide	0.02 *	0.1 (P), (1 D)	
	Methamidophos	Insecticide, Acaricide	0.01 *	—	[44]
	Diazinon	Insecticide, Acaricide	0.05	0.5 (D)	
	Dimethoate	Insecticide, Acaricide	0.01 *	3 (D)	
	Parathion—methyl	Insecticide	0.01 *	—	
	Chlorpyrifos	Insecticide	0.01 *	20 (D)	
Pepper	Malathion	Insecticide, Acaricide	0.02 *	0.1 (P), (1 D)	
	Azoxystrobin	Fungicide	3	3 (F), 30 (D)	[45]
	Cyprodinil	Fungicide	1.5	2 (F), (9, D)	
	Fludioxonil	Fungicide	1	1 (P), (4 D)	
	Lufenuron	Insecticide, Acaricide	0.8	—	
Pepper	Chlorothalonil	Fungicide	0.01 *	7 (P), (70, D),	[42]
	Endosulfan	Insecticide, Acaricide	0.05 *	—	
	Ethoprophos	Nematicide, Insecticide	0.05	0.2 (D)	
	Kresoxim—methyl	Fungicide	0.8	—	
	Pirimiphos—methyl	Insecticide, Acaricide	0.01 *	—	
Chili pepper	Procymidone	Fungicide	0.01 *	—	
	Aldrin	Insecticide	0.01 *	—	[46]
	Dieldrin	Insecticide	0.01 *	—	
	Endrin	Insecticide	0.01 *	—	
	HCB	Fungicide	0.01 *	—	
	Heptachlor	Insecticide	0.01 *	—	
	o,p′—DDT	Insecticide	0.05 *	—	
	p,p′—DDT	Insecticide	0.05 *	—	
Sweet pepper, Bell pepper	Chlorfenapyr	Insecticide, Acaricide	0.01 *	0.3 (P), (3 D)	[47]
	Triadimefon	Fungicide	0.01 *	1 (F), (5 D)	

(*) Indicates lower limit of analytical determination. (P) Group: Peppers (F) Group: FVOTC (D) Dry Pepper. FVOTC: Fruiting Vegetables, other than Cucurbits. EU: European Union.

3. Sample Pretreatment and Extraction Methods

Analytical methods are vital for estimating MRLs, from sample homogeneity to instrument detection limits. In pesticide research, substantial efforts have been undertaken to create and evaluate analytical techniques and procedures. Suppose the experimental sample is too small to accurately represent the initial batch or unit. In this case, applying sophisticated analytical tools and procedures would be expensive, time-consuming, and inefficient and could provide data that are challenging to understand instead of useful findings [48,49]. Consequently, efficient sample preparation is essential for accurately determining pesticide residues in foods with complex matrices [50].

The distribution of pesticide residues in/on crops is diverse. Thus, the sample needs to be completely homogenized. The matrix components are frequently coextracted with specific pesticides after obtaining a suitably homogeneous sample. Notably, more than 2500 natural compounds are found in paprika [51], which may hide the detection of some pesticide residues. In most conventional methods, the samples are extracted with acetonitrile and/or acetone. NaCl was added to the aqueous phase (either as a saturated solution or in solid form) to broaden the polarity range. Afterward, the extract was partitioned with nonpolar solutions (dichloromethane [DCM] or DCM/petroleum ether) to eliminate water and coextracts (e.g., pigments, phenols, and tannins obtained during liquid–liquid partitioning). The utilization of DCM in the liquid–liquid partitioning process was prohibited in 1980 because of the harmful impacts of chlorinated solvents on the environment and human health [52]. Therefore, many attempts have been made to substitute DCM or remove the liquid–liquid partitioning phase. In this context, a cyclohexane/ethyl acetate combination (1:1, v/v) was employed instead of DCM/petroleum ether (1:1, v/v) during the partitioning step [53,54]. Moreover, a solid-phase extraction (SPE) approach was used in place of liquid–liquid partitioning with DCM. For instance, Luke et al. [55] added fructose, $MgSO_4$, and NaCl to the original extract to separate the water from the acetone. To phase-separate mixtures of acetone/water and acetonitrile/water, Schenck et al. [54] used Na_2SO_4 and $MgSO_4$ as drying agents. The authors discovered that acetonitrile was more successfully and efficiently separated from the water than acetone and that $MgSO_4$ was more efficient in removing any remaining water from the organic layer.

Compared to other traditional techniques, the QuEChERS (quick, easy, cheap, effective, rugged, safe) method is widely used because of its many advantages, such as the ease of sample preparation, inexpensiveness, less organic solvent use, high recovery, and accuracy [56,57]. Noh et al. [58] described the QuEChERS technique as a streamlined strategy for analytical chemists to define the concentrations of multiclass and multi-residue pesticides in fruits and vegetables. With this method, $MgSO_4$ was used in a new way for salting-out extraction and partitioning with acetonitrile, cleaning with dispersive solid-phase extraction (d-SPE), and detection with mass spectrometry (MS). The initial version of the QuEChERS approach demonstrated remarkable performance in detecting hundreds of pesticides in various products. Nevertheless, using the initial approach caused the poor recovery of some pH-dependent pesticides, including pymetrozine, thiabendazole, and Imazalil [59]. This approach has undergone some modifications, mainly concerning pH variations and the use of a rather powerful acetate-buffered version, which became the official method of the Association of Official Analytical Chemists (AOAC) [43]. Instead, a citrate-buffered version was adopted as the European Standard (EN) procedure by the European Committee for Standardization (CEN) [60]. Due to frequent modifications of the solvents, salts, buffers, and sorbents used in the QuEChERS analytical approach, the QuEChERS approach is viewed as a sample preparation idea instead of a specific procedure [61]. Modifications are required to avoid pesticide degradation, achieve a reasonable recovery within an acceptable range, and lessen the matrix influence in complex matrices [62].

Challenges in Sample Preparation

Several challenges can arise during the sample preparation process to determine pesticide residues in peppers. Some of these challenges include the following:

- *Sample size*: Peppers can vary significantly in size, and can be challenging to accurately sample the fruit in a way that represents the overall population.
- *Contamination*: It is crucial to avoid the contamination of the sample during the preparation process, as this can affect the accuracy of the results.
- *Pesticide distribution*: Pesticides may not be uniformly distributed on the surface of the peppers, making it difficult to sample the fruit accurately.

- *Extraction efficiency:* The efficiency of the extraction process can impact the accuracy of the results, as some pesticides may be more difficult to extract than others.
- *Matrix effects:* The presence of other compounds in the peppers (such as proteins, carbohydrates, and lipids) can interfere with the analysis process and impact the accuracy of the results.

To address these challenges, it is vital to use appropriate sampling techniques and sample preparation methods and to carefully control the conditions of the analysis process to ensure the accuracy and reliability of results.

4. Cleanup Procedures

Before instrumentation, samples are often purified with sorbents, such as $MgSO_4$ combined with primary secondary amine (PSA), octadecylsilyl-derivatized silica (C18), and graphitized carbon black (GCB) [60,63]. The limited recovery of C18 in the analysis of nonpolar molecules and the great affinity of the GCB Table for planar analytes are two drawbacks of these often-employed sorbents. Therefore, additional efforts are needed to create novel sorbents or to optimize sorbent combinations to improve the purification effectiveness of matrices.

SPE was created to replace traditional partitioning and reduce the dangerous chlorinated solvents used in the partitioning stage [64]. The technique still needs a sizable glass column with sizable amounts of solvent for washing and elution, even though SPE was used in place of partitioning. Consequently, steps were taken to limit the consumption of solvents. The initial strategy used short florisil columns [64]. Instead of the large classical cartridges, C18 and Florisil column cartridges were also assessed in the cleanup of organo-halogen pesticides in crop matrices, and both cartridges showed acceptable recovery rates [65]. Therefore, SPE cartridges with normal or reversed-phase supports are now commercially available and provide a simple means for sample cleanup without requiring large amounts of solvent. Another extraction and cleanup method, known as matrix solid-phase dispersion (MSPD), was created to overcome the general limitations of liquid–liquid partitioning and SPE columns, including the requirement for many solvents and the emulsification of some fruits and vegetables, which blocks the flow of the analytes [66,67]. The MSPD strategy entails mixing a tiny quantity of the matrix with C18, washing it with a small amount of solvent, and eluting it to extract various chemicals (Barker, 2000a). Nonetheless, because of the minute sample size (0.5 g) used in this method, MSPD did not offer an analytical scope or a process that was adequately broad or straightforward. Anastassiades et al. [53] introduced d-SPE QuEChERS following a similar MSPD strategy [66,68,69]. The sorbent was then combined with an aliquot of the extract instead of the original sample, as in MSPD.

5. Instrumentation

Gas chromatography (GC), high-performance liquid chromatography (HPLC), and chromatography—mass spectrometry (GC–MS) are among the main methods that are used for pesticide residue detection and metabolite detection [11,28]. These conventional detection methods have good sensitivity, accuracy, precision, and reliability. However, they have some disadvantages, such as cumbersome sample pretreatment steps, the high cost of instruments and equipment requiring professional and technical personnel to operate them, and long detection processes. In this context, Rahman et al. [28] analyzed alachlor residues in the pepper and pepper leaves by GC and verified them through MS with pepper leaf matrix protection. They found that alachlor residues were present in both pepper and pepper leaf samples, with levels exceeding the MRL set by the Malaysian Food Regulation. The authors concluded that the consumption of peppers containing alachlor residues could pose a potential health risk to consumers. The study also highlighted the importance of the regular monitoring of pesticide residues in vegetables and fruits, as well as the need for stricter regulations on the use of pesticides in agriculture.

The main challenges in creating efficient methods for pesticide residue analysis include the low detection thresholds demanded by regulatory agencies, the variance in the polarity, volatility, and solubility of pesticides, and matrix coextraction [70]. Therefore, mass spectrometers, as a universal and more specific type of detector, began to be paired with chromatographic systems to overcome these problems [71]. In addition to improvements in the detection system, improvements in conventional sample preparation have been achieved regarding lowering the use of hazardous organic solvents, time, cost, and labor [72,73]. The QuEChERS sample preparation approach has met global acceptance and has been modified and adapted for various purposes due to its simplicity and flexibility [63,74]. However, this technique was created for gas chromatography–mass selective detection (GC–MSD) or liquid chromatography with tandem mass spectrometry (LC–MS/MS), which demand equipment that is uncommon in laboratories with only the most basic equipment [59,74].

In the quickly expanding food sector, automation in the analytical field is becoming increasingly important. Globally, strict rules and residue monitoring procedures are being created in response to consumer concerns about food safety. Due to the increased sample loads, high-throughput analytical techniques with sufficient precision and accuracy are needed.

6. Monitoring

The complexity of sample treatment largely depends on the matrix interferences and separation techniques, with GC and HPLC being the most common methods. Peppers typically have higher pesticide residue concentrations than other products because these compounds are constantly applied throughout the growing season. The research conducted in 2017 evaluated the levels of organochlorine pesticides in Nigerian noodles. The findings revealed that the chili peppers used in the noodles contained elevated levels of pesticide residues [41]. Several technologies have contributed to advancing the detection and monitoring of trace pesticide levels, including the GC-electron capture detector (ECD) [24,39,41,46] nitrogen phosphorus detector (NPD) [24,44], flame photometric detector (FPD) [38,43], GC–MS [30,35–37], and gas chromatography–tandem mass spectrometry (GC–MS/MS) due to the high selectivity, separation power, and identification capacity of MS (Table 2). The latest progress in MRMs associated with GC–MS/MS included the development of an analytical procedure that replaces traditional GC detectors. Nevertheless, due to the inadequate sensitivity for a few compounds, traditional GC detectors are still in use for SRMs [75].

To overcome inference problems, liquid chromatography with ultraviolet absorbance detection (LC-UVD) is commonly used [24,48,75]. Moreover, LC coupled with fluorescence detection (FLD) has been proposed as a promising solution to address suppression problems [42]. The need for cleanup has been decreased or eliminated. The method has been simplified, making it possible to recover all analytes in many different matrices via a single extraction and to detect them with either GC–MS/MS [75] or LC–MS/MS and ultrahigh-performance liquid chromatography–tandem mass spectrometry (UHPLC–MS/MS) [27,29,41,75].

Table 2. Overview of international monitoring for the analysis of pesticides in peppers.

Country	Number of Samples	Matrix	Number of Pesticides Detected	Sample Treatment	Determination Technique	Reference
China	299	Bell peppers	25 (15 OPs. 7 Pys. 3 CBs) 86. N > LOQ	Sample treatment 10 g sample + 25 mL MeCN → centrifugation (5000 rpm 5 min) → 3 g Nall vortex, upper layer sodium sulfate acetone/DCM (1:1; v/v) elution 15 mL	GC–MS	[30]
Turkey	725	Green pepper	170. 12.9% N > MRL N > LOQ	QuEChERS (Association of Official Analytical Chemists [AOAC]) (minor modification) 15 g homogenized sample + 15 mL MeOH/acetic acid (99:1, v/v) + 6 g anhydrous $MgSO_4$ + 1.5 g anhydrous sodium acetate, followed by vigorous shaking Centrifugation: Reconstitution in 8 mL upper MeOH + 900 mg anhydrous $MgSO_4$ + 150 mg PSA → dilution in 800 mL H_2O/MeOH (95:5, v/v) + 2 mmol/L ammonium formate	LC–MS/MS	[31]
Jordan	27.11	Sweet pepper/ Pepper	113, N > MRL. 13/4	QuEChERS 10 g sample + 10 mL MeCN 1 min vigorous shaking. Centrifugation: 0.5 mL supernatant + dilution 0.5 MeCN → 10 mL formic acid (5% in MeCN) final concentration ca. 1 g/mL	LC–MS/MS	[32]
Saudi Arabia	211	Chili pepper	80. N > 100.28 N > MRL. 14	QuEChERS 15 g sample + 15 mL MeCN + 1% acetic acid + 6 g $MgSO_4$ + 2.5 g sodium acetate trihydrate → shaking → centrifugation (4 min), 5 mL supernatant + 750 mg $MgSO_4$ + 250 mg PSA + graphitized carbon → shaking (20 s) → centrifugation (4000 rpm, 5 min)	UHPLC– MS/MS GC–MS/MS	[33]
EU	91.015	Sweet peppers/ Bell peppers	821. N > MRL 2.7%	EFSA (European Food Safety Authority)		[47]

Table 2. *Cont.*

Country	Number of Samples	Matrix	Number of Pesticides Detected	Sample Treatment	Determination Technique	Reference
Cameroon	6 2	Chili pepper/ White pepper	198. N > MRL. 38 Chili (23.2) White (20.21)	QuEChERS 5 g powder + 5 mL Milli-Q water + 15 mL MeCN 1 min vigorous shaking. A mixture of disodium hydrogen citrate sesquihydrate (0.75 g), trisodium citrate dihydrate (1.5 g), NaCl (1.5 g), and anhydrous $MgSO_4$ (16 g) was agitated (3 min) on a shaker (300 rpm) → centrifugation (10,000 rpm, 5 min) 8 mL supernatant + 300 mg PSA + 900 mg $MgSO_4$ + 150 mg Cl8 → shaking (1 min) → centrifugation (3000 rpm, 5 min) 6 mL supernatant → LC–MS/MS 5 mL supernatant → evaporation → replaced by 5 mL hexane + 2 mL extract → GC-ECD	LC–MS/MS GC-ECD	[76]
Poland	16	Pepper	242. N > LOD. 4.	Regulation of the Ministry of Agriculture and Rural Development from 27 Nov. 2013. (DL. U.z 2013 r. NR oo, Pol, 1549) A multi-residue chromatographic method was based on residue extraction with an organic solvent and the further purification of the extract with column chromatography. The final determination of residues was performed on gas chromatographs Agilent 7890 and Agilent 6890 equipped with electron capture (ECD) and nitrogen–phosphorus detectors (NPD). Dithiocarbamate fungicides were analyzed by a spectrophotometric method based on their decomposition to CS2 in the acid environment and transfer to methyl blue, which was then analyzed with the spectrometer Unicam Helios		[34]
Serbia	3	Pepper	6. N > MEL	SPME 10 g sample + 20 mL MeOH centrifugation → supernatant volume adjusted to 100 mL with water → 10 times dilution; the extraction times varied from 30 to 60 min	GC–MS	[35]

Table 2. *Cont.*

Country	Number of Samples	Matrix	Number of Pesticides Detected	Sample Treatment	Determination Technique	Reference
Botswana	83	Green pepper	232. N > MRL2 N > LOD 5	The official AOAC method 15 g sample + internal standard (90 mL) → shaking (1 min) → 15 mL 1% acetic acid in MeCN + extraction salt → vigorous shaking (1 min); centrifugation (3000 rcf, 5 min, 10 °C); cleanup → 1 mL supernatant + 2 mL DSPE tube → vortex (1 min) → centrifugation (3000 rcf, 5 min). 1 mL aliquot → GC–MS/MS (packed with 400 mg styrene-divinylbenzene copolymer LiChrolut EN) SPE	GC–MS	[36]
Saudi Arabia	160	Green pepper	23. N > LOD7 N > MRL 7	10 g sample + 20 mL acetone → homogenization (2 min); centrifugation (3000 rpm, 5 min); extraction columns conditioned with 6 mL ethyl acetate + 6 mL MeOH → 8 mL deionized H_2O; sample loading under vacuum at 5 mL/min flow rate → vacuum for 30 min, after which the pesticides were eluted into 3 × 2 mL aliquots of ethyl acetate/acetone; the eluate was evaporated to less than 1 mL (nitrogen) (90:10, v/v); solvent exchanged to 2 mL acetone → 6 mL acetone SPE	GC–MS	[37]
South Korea	7 8	Dried pepper leaves/ Dried red pepper	253. N > LOD 6 N > MRL 1 N > LOD 2	20 g sample + 50 mL H_2O + 100 mL MeCN → homogenization (2 min) 20 g anhydrous NaCl + extract → vortex (3 min) → centrifugation (4000 rpm, 5 min, 4 °C); 10 mL aliquot → evaporation (35–40 °C) (nitrogen); the mixture was passed through a Florisil cartridge (conditioned with 5 mL acetone/hexane (2:8, v/v). The solvent in the mixture was evaporated → 2 mL acetone → GC, GC–MS. The mixture was passed through SPE NH_2 cartridges (1 g, 6 mL). Cartridge treated with 5 mL DCM/MeOH (8:2, v/v). The solvent in the mixture was evaporated → 2 mL MeOH → HPLC-UV	GC-NPD/ECD GC–MSD HPLC-UV	[24]

Table 2. *Cont.*

Country	Number of Samples	Matrix	Number of Pesticides Detected	Sample Treatment	Determination Technique	Reference
Canada	90	Green pepper	N > LOD. 316	Column chromatography partitioning 50 g sample + 250 mL MeCN/H_2O (2:1, *v/v*) → filtration (one-half dilution with 500 mL water + 25 mL saturated NaCl) → partitioning with 2 × 50 mL DCM. The DCM was then dried by percolation through anhydrous Na_2SO_4 → isooctane was added → the sample was dried → redissolved in 5 mL isooctane, 4–5 mL concentrated sample cleaned up by Florisil column	GC-NPD/FPD	[38]
Spain	50	Pepper	31 N > LOD 35%918)	10 g sample + 50 mL ethyl acetate → homogenization (2 min) → 20 g sodium sulfate added → homogenization (1 min) → filtration (12 × 2 mL ethyl acetate) → evaporation → redissolved in 5 mL MeOH → LC–MS/MS	LC–MS/MS	[77]
China	15	Green pepper	33 N > LOD 4 N > MRL 1	SPE 50 g sample + 100 mL MeCN → oscillation (30 mL) → + 5 g NaCl → centrifugation (3000 rpm, 5 min) → MeCN layer evaporation (to 1 mL) → filtration using SPE column (Carb/NH_2) → ENVI-Carb (6 mL/500 mg) +NH_2-LC (6 mL/500 mg) + anhydrous sodium sulfate → wash with 5 mL acetone/methylbenzene (1:1, *v/v*) → wash four times with MeCN→ eluent evaporation → dissolve with MeCN → evaporation to 0.5 mL (nitrogen stream, 80 °C) → eluant was quantified to MeCN (1 mL)	GC-ECD/FPD	[39]
Cameroon	11	Chili pepper	20 N > LCD 58.9%	QuEChERS	GC-ECD	[46]
India	4	Chili pepper	8 N > LOD 4	Blend + extract with MeCN → re-extract with *n*-hexane layer by partitioning process → pigments removed by activated charcoal → extracts cleaned with concentrated H_2SO_4	GC-ECD	[78]
Slovenia	21	Pepper	214 N > LOQ 19 N > MRL 2	Cleanup by gel permeation chromatography → GC−MS sample heated two-phase system isooctane/stannous(II) chloride in diluted HCl; carbon disulfide dissolved in the organic phase (isooctane) → GC−MS	LC–MS/MS	[45]

Table 2. *Cont.*

Country	Number of Samples	Matrix	Number of Pesticides Detected	Sample Treatment	Determination Technique	Reference
Venezuela	16	Red pepper	7 N > LOQ 14 N > MRL 6	200 g sample, chopped subsamples were weighed (4 g) in triplicate + 10 mL ethyl acetate/acetone (90:10, v/v) + 5 g anhydrous sodium sulfate. The solvent (organic layer) concentration was adjusted to 20 mL → GC-NPD	GC-NPD	[44]
Ghana	50	Pepper	9 N > LOQ 8 N > MRL 8	2 g sample + 5 mL MeCN → agitation (15 min) → MeCN layer decantation, MeCN layer centrifugation (3000 rpm, 2 min) → concentration to 2 mL with MeCN → GC-FPD	GC-FPD	[43]
Republic of Korea	1207	Pepper	250 N > LOQ N > MRL 12 (2003.8.5 2004.12.0 2005. 13.3%)	50 g sample + 100 mL MeCN (2 min) → homogenization → 10–15 NaCl → extract shaken vigorously → stand (30 min); 10 mL upper phase evaporation to dryness (60 °C, air stream) → next three cleanup steps repeated three times. (1) Dissolved in acetone (2 mL) → filtered through a 0.2 μm nylon Acrodisc (Whatman) → GC-NPD. MS (2) Dissolved in 20% acetone/hexane (2 mL) and loaded onto a Sep-Pak Florisil cartridge (Phenomenex) preconditioned with hexane (5 mL), → 20% acetone in hexane (5 mL). The cartridge was eluted with 20% acetone/hexane (5 mL) twice → evaporation → dissolved in 20% acetone in hexane (2 mL) → GC-ECD. (3) Dissolved in 1% MeOH/methylene chloride (2 mL) → loaded onto a Sep-Pak NH_2 cartridge (Varian) → reconditioned with 1% MeOH/methylene chloride (5 mL) twice → elution evaporation → dissolved in MeOH (5 mL) → a total of 2 mL of the 5 mL MeOH → LC-DAD, MS. Total of 3 mL of the 5 mL MeOH + 2 mL of 1% acetic acid (pH 3) → filtration (0.2 μm nylon Acrodisc) → LC-FLD	GC–MSD LC-FLD LC–MSD	[42]

Table 2. *Cont.*

Country	Number of Samples	Matrix	Number of Pesticides Detected	Sample Treatment	Determination Technique	Reference
Nigeria	12	Chili pepper/ Green bell pepper	13	Ultrasonic extraction 5 g sample + 2.5 g anhydrous sodium sulfate +20 mL ethyl acetate → shaking (270 rpm, 5 min) → sonication (40 °C, 20 min) → stand (5 min) → centrifugation (2500 rpm, 5 min) → supernatant concentration (1 mL, nitrogen gas) → SPE cartridge (conditioning MeOH, water, ethyl acetate) → sample extract loaded onto SPE → elution n-hexane/DCM (3:2, v/v) → eluate concentration (1 mL n-hexane, amber vial) → GC-ECD	GC-ECD	[41]
Mexico	207	Chili pepper source	1 (ƆPE)	European Environment Agency (EPA, 1981)	GC	[79]

7. Effect of Household Processing on Pesticide Residue Levels in Peppers

Several household processes can be used to reduce pesticides in fresh peppers, including washing and blanching. These processes can effectively remove or reduce the levels of pesticides on the surface of peppers, but they may not completely eliminate all residues. For instance, washing peppers thoroughly using running water can effectively remove surface contaminants, including pesticides, but it will not remove all pesticides, particularly those that have been absorbed into the pepper tissue [13]. Blanching is a process in which peppers are briefly boiled in water or steam and then cooled in ice water. This process can help loosen the pepper's skin, making it easier to remove. It can also help to reduce the levels of pesticides on the surface of the pepper, as some pesticides may be removed during the boiling process [80]. Again, blanching may not be able to remove all pesticides, particularly those that have been absorbed into the pepper tissue. In this context, Kim et al. [80] evaluated the effects of various household processes, such as washing, blanching, frying, and drying, under different conditions (water volume, blanching time, and temperature) on residual pesticide concentrations. Both washing and blanching (in combination with high water volume and processing time) significantly reduced pesticide residue levels in the leaves and fruit of hot pepper compared with other processes [80]. It is worth considering other conditions/factors, such as selecting peppers that are grown using sustainable and organic practices to further reduce the levels of pesticides in peppers.

8. Dissipation Patterns and Preharvest Intervals in Peppers

Pesticides that are applied to peppers can be absorbed by the plant while also being present on the surface of the pepper fruit. The rate at which pesticides dissipate or break down can vary depending on several factors, including the type of pesticide, the application rate, the weather, and the application method. Generally, most pesticides will dissipate more quickly in warm, humid conditions and more slowly in cool, dry conditions. Pesticides applied to the surface of the pepper fruit may dissipate more quickly than those absorbed by the plant, as they are more exposed to the environment.

The dissipation behavior of pesticide residues in peppers has been investigated [81–87]. For instance, Liu et al. [85] reported that the $t_{1/2}$ values of metalaxyl in peppers were 3.2–3.9 days at three experimental locations in China. At harvest, pepper samples were found to contain metalaxyl and cymoxanil levels that were well below the MRLs of the EU following the recommended dosage and an interval of 21 days after the last application.

The environmental fate of field-applied synthetic pesticides has been under investigation for several years. Endosulfan 3 EC, a mixture of α- and β-stereoisomers, was sprayed on field-grown pepper at the recommended rate of 0.44 kg of active ingredients per acre. Endosulfan sulfate is the major metabolite of endosulfan sulfite, and the β-isomers are relatively more persistent than the α-isomers. In pepper, the α-isomer, which is more toxic to mammals, dissipated faster ($t_{1/2}$ = 1.22 day) than the less toxic β-isomer ($t_{1/2}$ = 3.0 day). These results confirm the greater loss of the α-isomer than the β-isomer, which can ultimately impact endosulfan dissipation in the environment [82].

The degradation behavior of flonicamid and its metabolites, 4-(trifluoromethyl)nicotinic acid (TFNA) and *N*-(4-trifluoromethylnicotinoyl) glycine (TFNG), was evaluated in red bell peppers over 90 days under greenhouse conditions, including high temperature, low and high humidity, and in a vinyl house covered with a high-density polyethylene light shade covering film (35% and 75%). For safety reasons, the authors concluded that red bell peppers should be grown under greenhouse conditions because solar radiation increases the rate of flonicamid degradation into its metabolites [88].

It is also possible to reduce the need for pesticides by using integrated pest management techniques, such as introducing natural predators of pests or using physical barriers to prevent pests from accessing plants.

PHIs are the minimum amount of time that must pass between the application of a pesticide and the harvest of a crop. The purpose of PHIs is to allow pesticides to break down or dissipate in the environment and on the surface of the crop to levels that are considered safe for consumption. PHIs vary depending on the specific pesticide used, the type of crop, and the application method. It is important to follow the label instructions for a particular pesticide, as these will include the recommended PHI for the crop in question to ensure that the peppers are safe to consume. It is also worth noting that some pesticides may not be approved for peppers, which means there would be no recommended PHI. It is important to use pesticides only as directed and to follow all label instructions to help ensure the safety of the crop and to protect human health.

9. Dietary Risk Assessment

The dietary risk assessment of pesticide residues in peppers is an important task that helps determine the potential health effects of consuming peppers treated with pesticides. This assessment typically involves several steps, including:

1. Identify the pesticides that are commonly used on peppers, as well as their maximum residue levels (MRLs).
2. Collect data on the levels of pesticide residues found in peppers sold on the market.
3. Evaluate the potential health risks posed by the consumption of peppers with pesticide residues based on the levels found and the MRLs.

Once the data are collected, they can be used to estimate the average daily intake of each pesticide for different population groups. This can be performed by using data on pepper consumption patterns and the levels of pesticide residues found in peppers. Next, the potential health risks posed by the consumption of peppers with pesticide residues can be evaluated by comparing the estimated daily intake of each pesticide with the appropriate reference doses (RfDs), such as acceptable daily intakes (ADIs) or acute reference doses (ARfDs) [30,89]. These values are established by regulatory agencies, such as the US Environmental Protection Agency (EPA), as a safe level of exposure for the general population. It is worth mentioning that to have a comprehensive view of the impact of pesticide residues in peppers on human health, it is crucial to look not only at the impact of a single pesticide but also at the combined effect of different pesticides that may be present in the pepper [90–92]. While the impact of individual pesticides on human health has been extensively studied, the combined effect of multiple pesticides is less understood. However, there is growing evidence to suggest that exposure to multiple pesticides can have additive or synergistic effects on human health and that the cumulative effect of these residues may be greater than the effect of individual pesticides alone. Therefore, it is important to consider the potential combined impact of multiple pesticide residues when evaluating the health risks associated with consuming peppers or other fruits and vegetables. In addition, the levels of the detected pesticide in peppers can be tolerated and do not pose a serious health problem to the community [30,31]. However, it is worth noting that some people may be more sensitive to pesticides than others, such as pregnant women and children [93]. Additionally, long-term exposure to low levels of pesticides may also pose health risks [94]. It is also important to note that the risk assessment process may vary by country, as different countries have different regulations for pesticides, different exposure scenarios, and different methods for assessing risks. It is worth mentioning that regulatory agencies continuously monitor the situation and update their guidelines and regulations as necessary.

10. The Use of Pepper Leaf Matrix as an Analyte Protectant

The pepper leaf matrix is a complex mixture of compounds that are found in the leaves of pepper plants. It is composed of various organic compounds, such as proteins, carbohydrates, and lipids, as well as inorganic compounds, such as minerals. The specific composition of the pepper leaf matrix depends on the pepper plant variety and the growing conditions. It is possible that the pepper leaf matrix could be used as an analyte protectant during GC analysis [26,28]. Analyte protectants are substances used to stabilize or protect specific molecules or compounds during the analysis process. This can help prevent the degradation or loss of the analyte [26,28], ensuring that accurate and reliable results are obtained.

11. Challenges and Limitations in Managing Pesticide Residues in Peppers: Author's Perspectives

There are several challenges and limitations when measuring and managing pesticide residues in peppers. Some of the main challenges and limitations include the following:

- *Detection limits*: Many pesticides break down or degrade over time, making it difficult to accurately measure their residues in peppers. This can be incredibly challenging when trying to detect low levels of pesticides, as the limits of detection for many analytical methods may be higher than the levels of residues present in the peppers.
- *Matrix interference*: The presence of other substances in the pepper sample, such as sugars and other organic compounds, can interfere with the accuracy of pesticide residue analysis.
- *Sample preparation*: Preparing samples for pesticide residue analysis can be time-consuming and labor-intensive. It is important to follow proper sample preparation procedures to ensure that the samples are representative and that the analysis results are accurate.
- *Regulatory limits*: Different countries and regions have different regulations and guidelines for the MRLs of pesticides in peppers and other food products. Ensuring that peppers meet these regulatory limits can be challenging, especially when dealing with multiple pesticides and different regulatory frameworks.
- *Pesticide resistance*: Some pests and diseases that affect peppers can develop resistance to certain pesticides over time. This can make it more challenging to control these pests and diseases and can lead to the need for more frequent or higher applications of pesticides.

Overall, managing pesticide residues in peppers can be a complex and challenging task. It is important to follow proper pesticide application and management practices to minimize the levels of residues in peppers and to ensure that they meet regulatory limits.

12. Conclusions and Future Perspectives

Depending on their use, pesticides might have positive and negative effects on peppers. It is important to use pesticides responsibly and to follow all label instructions to minimize any potential negative effects on peppers and other non-target organisms. The analytical approach to monitoring pesticide residues in peppers and their monitoring frequency is important in ensuring the safety of the food supply. Various analytical methods and sample preparation techniques are available, and regulatory agencies and food safety organizations play a crucial role in monitoring pesticide residues in peppers to ensure that they are safe for human consumption. Notably, managing pesticide residues in peppers can be a complex and challenging task. Therefore, it is essential to follow proper pesticide application and management practices to minimize the levels of residues and ensure that they meet regulatory limits. A trend toward using safer and more sustainable pest control methods in pepper production, appropriate sample cleanup methods, techniques designed to remove matrix interferences (such as pigments, lipids, and carbohydrates) and purify the target analytes effectively, and the use of accurate and reliable analytical methods should be considered. It is also important to wash and peel peppers thoroughly before consuming them to reduce the risk of exposure to pesticide residues. Overall, the outcomes of pesticide residue analysis in peppers depend on the specific method used, the type and concentration of pesticides detected, and the regulatory standards that apply.

There are several potential future developments in pesticide residue analysis in peppers that may emerge in the coming years. For instance, more sensitive and accurate methods could be developed to detect the trace levels of new pesticides. As consumer demand for organic produce grows, there may be an increased focus on alternative pest control methods that do not involve synthetic pesticides. This could lead to a decrease in the levels of pesticide residues that are found in peppers. Automating analytical techniques could also become more widespread in the future, improving the efficiency and accuracy of pesticide residue analysis in peppers. MRLs for pesticides in food, including peppers, are periodically reviewed and updated. There is an ongoing debate about what levels of pesticide residues are safe for human consumption and how MRLs should be established. Risk assessment methods are also being developed to help determine the potential health risks associated with different levels of pesticide residues in peppers.

Author Contributions: J.-H.S.: conceptualization, resources, validation, writing—review and editing. J.-B.E.: data curation. A.A.Z.: validation, data curation, writing—review and editing. A.S.H. and A.H.: validation. A.M.A.: visualization, data curation, writing—review and editing. All authors have read and agreed to the published version of the manuscript.

Funding: This research received no external funding.

Data Availability Statement: Data is contained within the article.

Conflicts of Interest: The authors declare no conflict of interest.

References

1. Kim, I.K.; Abd El-Aty, A.M.; Shin, H.C.; Lee, H.B.; Kim, I.S.; Shim, J.H. Analysis of volatile compounds in fresh healthy and diseased peppers (*Capsicum annuum* L.) using solvent free solid injection coupled with gas chromatography-flame ionization detector and confirmation with mass spectrometry. *J. Pharm. Biomed. Anal.* **2007**, *45*, 487–494. [CrossRef] [PubMed]
2. Sanatombi, K.; Rajkumari, S. Effect of processing on quality of pepper: A review. *Food Rev. Int.* **2020**, *36*, 626–643. [CrossRef]
3. Park, S.-Y.; Kim, J.-Y.; Lee, S.-M.; Jun, C.-H.; Cho, S.-B.; Park, C.-H.; Joo, Y.-E.; Kim, H.-S.; Choi, S.-K.; Rew, J.-S. Capsaicin induces apoptosis and modulates MAPK signaling in human gastric cancer cells. *Mol. Med. Rep.* **2014**, *9*, 499–502. [CrossRef] [PubMed]
4. Dreher, M.L.; Davenport, A.J. Hass avocado composition and potential health effects. *Crit. Rev. Food Sci. Nutr.* **2013**, *53*, 738–750. [CrossRef]
5. Sousa, E.T.; Rodrigues FD, M.; Martins, C.C.; de Oliveira, F.S.; Pereira, P.A.D.P.; de Andrade, J.B. Multivariate optimization and HS-SPME/GC–MS analysis of VOCs in red, yellow and purple varieties of *Capsicum chinense* sp. peppers. *Microchem. J.* **2006**, *82*, 142–149. [CrossRef]
6. Kim, S.; Park, J.; Hwang, I.K. Composition of main carotenoids in Korean red pepper (*Capsicum annuum* L.) and changes of pigment stability during the drying and storage process. *J. Food Sci.* **2004**, *69*, FCT39–FCT44. [CrossRef]
7. Fulton, J.C.; Uchanski, M.E. A review of chile pepper (*Capsicum annuum*) stip: A physiological disorder of peppers. *HortScience* **2017**, *52*, 4–9. [CrossRef]
8. García-Gaytán, V.; Gómez-Merino, F.C.; Trejo-Téllez, L.I.; Baca-Castillo, G.A.; García-Morales, S. The chilhuacle chili (*Capsicum annuum* L.) in Mexico: Description of the variety, its cultivation, and uses. *Int. J. Agron.* **2017**, *2017*, 5641680. [CrossRef]
9. Food and Agriculture Organization of the United Nations (FAO). FAOSTAT. 2021. Available online: http://www.fao.org/faostat/en/#data/QC (accessed on 26 November 2022).
10. Lee, M.G.; Jung, D.I. Processing factors and removal ratios of select pesticides in hot pepper leaves by a successive process of washing, blanching, and drying. *Food Sci. Biotechnol.* **2009**, *18*, 1076–1082.
11. Shim, J.H.; Rahman, M.M.; Zaky, A.A.; Lee, S.J.; Jo, A.; Yun, S.H.; Eun, J.-B.; Kim, J.-H.; Park, J.-W.; Oz, E.; et al. Simultaneous determination of pyridate, quizalofop-ethyl, and cyhalofop-butyl residues in agricultural products using liquid chromatography-tandem mass spectrometry. *Foods* **2022**, *11*, 899. [CrossRef]
12. Oerke, E.C.; Dehne, H.W. Safeguarding production—Losses in major crops and the role of crop protection. *Crop Prot.* **2004**, *23*, 275–285. [CrossRef]
13. Radwan, M.A.; Abu-Elamayem, M.M.; Shiboob, M.H.; Abdel-Aal, A. Residual behavior of profenofos on some field-grown vegetables and its removal using various washing solutions and household processing. *Food Chem. Toxicol.* **2005**, *43*, 553–557. [CrossRef]
14. WHO. Pesticide Poisoning and Public Health. 2017. Available online: https://www.who.int/whr/1997/media_centre/executive_summary1/en (accessed on 18 October 2022).
15. Narenderan, S.T.; Meyyanathan, S.N.; Babu, B.J.F.R.I. Review of pesticide residue analysis in fruits and vegetables. Pretreatment, extraction and detection techniques. *Food Res. Int.* **2020**, *133*, 109141. [CrossRef]
16. Pimentel, D. Environmental and economic costs of the application of pesticides primarily in the United States. *Environ. Dev. Sustain.* **2005**, *7*, 229–252. [CrossRef]
17. Keikotlhaile, B.M.; Spanoghe, P.; Steurbaut, W. Effects of food processing on pesticide residues in fruits and vegetables: A meta-analysis approach. *Food Chem. Toxicol.* **2010**, *48*, 1–6. [CrossRef]
18. Valverde, A.; Aguilera, A.; Rodríguez, M.; Boulaid, M.; Soussi-El Begrani, M. Pesticide residue levels in peppers grown in a greenhouse after multiple applications of pyridaben and tralomethrin. *J. Agric. Food Chem.* **2002**, *50*, 7303–7307. [CrossRef]

19. Ambrus, A.; Yang, Y.Z. Global harmonization of maximum residue limits for pesticides. *J. Agric. Food Chem.* **2016**, *64*, 30–35. [CrossRef]
20. CODEX. 2012. Available online: https://www.fao.org/fao-who-codexalimentarius/standards/pesters/commodities-detail/en/acid=239 (accessed on 11 December 2022).
21. Arias, L.A.; Bojacá, C.R.; Ahumada, D.A.; Schrevens, E. Monitoring of pesticide residues in tomato marketed in Bogota, Colombia. *Food Control* **2014**, *35*, 213–217. [CrossRef]
22. Bempah, C.K.; Buah-Kwofie, A.; Enimil, E.; Blewu, B.; Agyei-Martey, G. Residues of organochlorine pesticides in vegetables marketed in Greater Accra Region of Ghana. *Food Control* **2012**, *25*, 537–542. [CrossRef]
23. Kim, B.; Baek, M.S.; Lee, Y.; Paik, J.K.; Chang, M.I.; Rhee, G.S.; Ko, S. Estimation of apple intake for the exposure assessment of residual chemicals using Korea National Health and nutrition examination survey database. *Clin. Nutr. Res.* **2016**, *5*, 96–101. [CrossRef]
24. Seo, Y.H.; Cho, T.H.; Hong, C.K.; Kim, M.S.; Cho, S.J.; Park, W.H.; Hwang, I.S.; Kim, M.S. Monitoring and risk assessment of pesticide residues in commercially dried vegetables. *Prev. Nutr. Food Sci.* **2013**, *18*, 145–149. [CrossRef] [PubMed]
25. Yang, A.G.; Shim, K.H.; Choi, O.J.; Park, J.H.; Do, J.A.; Oh, J.H.; Shim, J.H. Establishment of the Korean total diet study (TDS) model in consideration to pesticide intake. *Korean J. Pestic. Sci.* **2012**, *16*, 151–162. [CrossRef]
26. Rahman, M.M.; Choi, J.H.; Abd El-Aty, A.M.; Abid, M.D.; Park, J.H.; Na, T.W.; Kim, Y.D.; Shim, J.H. Pepper leaf matrix as a promising analyte protectant prior to the analysis of thermolabile terbufos and its metabolites in pepper using GC–FPD. *Food Chem.* **2012**, *133*, 604–610. [CrossRef]
27. Rahman, M.M.; Choi, J.H.; Abd El-Aty, A.M.; Park, J.H.; Park, J.Y.; Im, G.J.; Shim, J.H. Determination of chlorfenapyr in leek grown under greenhouse conditions with GC-μECD and confirmation by mass spectrometry. *Biomed. Chromatogr.* **2012**, *26*, 172–177. [CrossRef] [PubMed]
28. Rahman, M.M.; Sharma, H.M.; Park, J.H.; Abd El-Aty, A.M.; Choi, J.H.; Nahar, N.; Shim, J.H. Determination of alachlor residues in pepper and pepper leaf using gas chromatography and confirmed via mass spectrometry with matrix protection. *Biomed. Chromatogr.* **2013**, *27*, 924–930. [CrossRef]
29. Siddamallaiah, L.; Mohapatra, S. Residue level and dissipation pattern of spiromesifen in cabbage and soil from 2-year field study. *Environ. Monit. Assess.* **2016**, *188*, 1–12. [CrossRef]
30. Chu, Z.; Zhuang, M.; Li, S.; Xiao, P.; Li, M.; Liu, D.; Zhou, J.; Chen, J.; Zhao, J. Residue levels and health risk of pesticide residues in bell pepper in Shandong. *Food Addit. Contam. Part A* **2019**, *36*, 1385–1392. [CrossRef]
31. Golge, O.; Hepsag, F.; Kabak, B. Health risk assessment of selected pesticide residues in green pepper and cucumber. *Food Chem. Toxicol.* **2018**, *121*, 51–64. [CrossRef]
32. Algharibeh, G.R.; AlFararjeh, M.S. Pesticide residues in fruits and vegetables in Jordan using liquid chromatography/tandem mass spectrometry. *Food Addit. Contam. Part B* **2019**, *12*, 65–73. [CrossRef]
33. Ramadan, M.F.; Abdel-Hamid, M.M.; Altorgoman, M.M.; AlGaramah, H.A.; Alawi, M.A.; Shati, A.A.; Shweeta, H.A.; Awwad, N.S. Evaluation of pesticide residues in vegetables from the Asir Region, Saudi Arabia. *Molecules* **2020**, *25*, 205. [CrossRef]
34. Slowik-Borowiec, M.; Szpyrka, E.; Rupar, J.; Podbielska, M.; Matyaszek, A. Occurrence of pesticide residues in fruiting vegetables from production farms in southeastern region of Poland. *Rocz. Państwowego Zakładu Hig.* **2016**, *67*, 359–365.
35. Marković, M.; Cupać, S.; Đurović, R.; Milinović, J.; Kljajić, P. Assessment of heavy metal and pesticide levels in soil and plant products from agricultural area of Belgrade, Serbia. *Arch. Environ. Contam. Toxicol.* **2010**, *58*, 341–351. [CrossRef]
36. Gondo, T.F.; Kamakama, M.; Oatametse, B.; Samu, T.; Bogopa, J.; Keikotlhaile, B.M. Pesticide residues in fruits and vegetables from the southern part of Botswana. *Food Addit. Contam. Part B* **2021**, *14*, 271–280. [CrossRef]
37. Osman, K.A.; Al-Humaid, A.M.; Al-Rehiayani, S.M.; Al-Redhaiman, K.N. Monitoring of pesticide residues in vegetables marketed in Al-Qassim region, Saudi Arabia. *Ecotoxicol. Environ. Saf.* **2010**, *73*, 1433–1439. [CrossRef]
38. Frank, R.; Braun, H.E.; Ripley, B.D. Residues of insecticides, and fungicides in fruit produced in Ontario, Canada, 1986–1988. *Food Addit. Contam.* **1990**, *7*, 637–648. [CrossRef]
39. Wang, Z.; Zhang, Y.; Wang, J.; Guo, R. Pesticide residues in market foods in Shaanxi Province of China in 2010. *Food Chem.* **2013**, *138*, 2016–2025. [CrossRef]
40. Benzidane, C.; Dahamna, S. Chlorpyrifos residues in food plant in the region of Setif-Algeria. *Commun. Agric. Appl. Biol. Sci.* **2013**, *78*, 157–160.
41. Oyeyiola, A.O.; Fatunsin, O.T.; Akanbi, L.M.; Fadahunsi, D.E.; Moshood, M.O. Human health risk of organochlorine pesticides in foods grown in Nigeria. *J. Health Pollut.* **2017**, *7*, 63–70. [CrossRef]
42. Cho, T.H.; Kim, B.S.; Jo, S.J.; Kang, H.G.; Choi, B.Y.; Kim, M.Y. Pesticide residue monitoring in Korean agricultural products, 2003–2005. *Food Addit. Contam. Part B* **2009**, *2*, 27–37. [CrossRef]
43. Darko, G.; Akoto, O. Dietary intake of organophosphorus pesticide residues through vegetables from Kumasi, Ghana. *Food Chem. Toxicol.* **2008**, *46*, 3703–3706. [CrossRef]
44. Quintero, A.; Caselles, M.J.; Ettiene, G.; De Colmenares, N.G.; Ramírez, T.; Medina, D. Monitoring of organophosphorus pesticide residues in vegetables of agricultural area in Venezuela. *Bull. Environ. Contam. Toxicol.* **2008**, *81*, 393–396. [CrossRef] [PubMed]
45. Galani, H.B.; Bolta, Š.V.; Gregorčič, A. Pesticide residues in cauliflower, eggplant, endive, lettuce, pepper, potato and wheat of the Slovene origin found in 2009. *Acta Chim. Slov.* **2010**, *57*, 972–979.

46. Galani YJ, H.; Houbraken, M.; Wumbei, A.; Djeugap, J.F.; Fotio, D.; Gong, Y.Y.; Spanoghe, P. Contamination of foods from Cameroon with residues of 20 halogenated pesticides, and health risk of adult human dietary exposure. *Int. J. Environ. Res. Public Health* **2021**, *18*, 5043. [CrossRef] [PubMed]

47. European Food Safety Authority (EFSA); Medina-Pastor, P.; Triacchini, G. The 2018 European Union report on pesticide residues in food. *EFSA J.* **2020**, *18*, e0605.

48. Farha, W.; Rahman, M.M.; Abd El-Aty, A.M.; Jung, D.I.; Kabir, M.H.; Choi, J.H.; Kim, S.W.; Jeong Im, S.; Lee, Y.J.; Shim, J.H. A combination of solid-phase extraction and dispersive solid-phase extraction effectively reduces the matrix interference in liquid chromatography–ultraviolet detection during pyraclostrobin analysis in perilla leaves. *Biomed. Chromatogr.* **2015**, *29*, 1932–1936. [CrossRef]

49. Lehotay, S.J.; Cook, J.M. Sampling and sample processing in pesticide residue analysis. *J. Agric. Food Chem.* **2015**, *63*, 4395–4404. [CrossRef]

50. Watanabe, E.; Kobara, Y.; Yogo, Y. Rapid and simple analysis of pesticides persisting on green pepper surfaces swabbing with solvent-moistened cotton. *J. Agric. Food Chem.* **2012**, *60*, 9000–9005. [CrossRef]

51. Vázquez, P.P.; Ferrer, C.; Bueno, M.M.; Fernández-Alba, A.R. Pesticide residues in spices and herbs: Sample preparation methods and determination by chromatographic techniques. *Trends Anal. Chem.* **2019**, *115*, 13–22. [CrossRef]

52. Seymour, J.; Korte, N.; Miles, C. *Handbook of Pesticide Toxicology*, 2nd ed.; Academic Press: Cambridge, MA, USA, 2003.

53. Anastassiades, M.; Lehotay, S.J.; Štajnbaher, D.; Schenck, F.J. Fast and easy multiresidue method employing acetonitrile extraction/partitioning and "dispersive solid-phase extraction" for the determination of pesticide residues in produce. *J. AOAC Int.* **2003**, *86*, 412–431. [CrossRef]

54. Schenck, F.J.; Callery, P.; Gannett, P.M.; Daft, J.R.; Lehotay, S.J. Comparison of magnesium sulfate and sodium sulfate for removal of water from pesticide extracts of foods. *J. AOAC Int.* **2002**, *85*, 1177–1180. [CrossRef]

55. Luke, M.A.; Cassias, I.; Yee, S. *Office of Regulatory Affairs*; Laboratory Information Bulletin No. 4178; US Food and Drug Administration: Rockville, MD, USA, 1999.

56. Jung, H.N.; Park, D.H.; Choi, Y.J.; Kang, S.H.; Cho, H.J.; Choi, J.M.; Shim, J.H.; Zaky, A.A.; Abd El-Aty, A.M.; Shin, H.C. Simultaneous quantification of chloramphenicol, thiamphenicol, florfenicol, and florfenicol amine in animal and aquaculture products using liquid chromatography-tandem mass spectrometry. *Front. Nutr.* **2022**, *8*, 1181. [CrossRef]

57. Tan, H.; Gu, Y.; Liu, S.; Zhang, H.; Li, X.; Zeng, D. Rapid residue determination of cyenopyrafen in citrus peel, pulp, and whole fruit using ultra-performance liquid chromatography/tandem mass spectrometry. *Food Anal. Methods* **2018**, *11*, 2123–2130. [CrossRef]

58. Noh, H.H.; Kim, C.J.; Kwon, H.; Kim, D.; Moon, B.-C.; Baek, S.; Oh, M.-S.; Kyung, K.S. Optimized residue analysis method for broflanilide and its metabolites in agricultural produce using the QuEChERS method and LC-MS/MS. *PLoS ONE* **2020**, *15*, e0235526. [CrossRef]

59. da Costa Morais, E.H.; Collins, C.H.; Jardim, I.C.S.F. Pesticide determination in sweet peppers using QuEChERS and LC–MS/MS. *Food Chem.* **2018**, *249*, 77–83. [CrossRef]

60. Anastassiades, M. Foods of Plant Origin–Determination of Pesticide Residues Using GC–MS and/or LCMS/MS Following Acetonitrile Extraction/Partitioning and Clean-up by Dispersive SPE (QuEChERS method. 2007). 2007. Available online: https://standards.iteh.ai/catalog/standards/cen/9f9e56e8-ac1c-4f3e-9f91-23d42/03dd8a/en-15662-2008 (accessed on 17 November 2022).

61. Rejczak, T.; Tuzimski, T. A review of recent developments and trends in the QuEChERS sample preparation approach. *Open Chem.* **2015**, *13*, 980–1010. [CrossRef]

62. González-Curbelo, M.Á.; Socas-Rodríguez, B.; Herrera-Herrera, A.V.; González-Sálamo, J.; Hernández-Borges, J.; Rodríguez-Delgado, M.Á. Evolution and applications of the QuEChERS method. *Trends Anal. Chem.* **2015**, *71*, 169–185. [CrossRef]

63. Lehotay, S.J.; Son, K.A.; Kwon, H.; Koesukwiwat, U.; Fu, W.; Mastovska, K.; Hoh, E.; Leepipatpiboon, N. Comparison of QuEChERS sample preparation methods for the analysis of pesticide residues in fruits and vegetables. *J. Chromatogr. A* **2010**, *1217*, 2548–2560. [CrossRef]

64. Das, S.K. Recent developments in clean up techniques of pesticide residue analysis for toxicology study: A critical review. *Univ. J. Agric. Res.* **2014**, *2*, 198–202. [CrossRef]

65. Rodríguez-Delgado, M.Á.; Hernández-Borges, J. Rapid analysis of triazolopyrimidine sulfoanilide herbicides in waters and soils by high-performance liquid chromatography with UV detection using a C18 monolithic column. *J. Sep. Sci.* **2007**, *30*, 8–14. [CrossRef]

66. Ferreira, J.A.; Santos LF, S.; Souza NR, D.S.; Navickiene, S.; de Oliveira, F.A.; Talamini, V. MSPD sample preparation approach for reversed-phase liquid chromatographic analysis of pesticide residues in stem of coconut palm. *Bull. Environ. Contam. Toxicol.* **2013**, *91*, 160–164. [CrossRef]

67. Purdešová, A.; Dömötorová, M. MSPD as sample preparation method for determination of selected pesticide residues in apples. *Acta Chim. Slovaca* **2017**, *10*, 41–46. [CrossRef]

68. Barker, S.A. Applications of matrix solid-phase dispersion in food analysis. *J. Chromatogr. A* **2000**, *880*, 63–68. [CrossRef] [PubMed]

69. Barker, S.A. Matrix solid-phase dispersion. *J. Chromatogr. A* **2000**, *885*, 115–127. [CrossRef] [PubMed]

70. Nguyen, T.D.; Lee, M.H.; Lee, G.H. Multiresidue determination of 156 pesticides in watermelon by dispersive solid phase extraction and gas chromatography/mass spectrometry. *Bull. Korean Chem. Soc.* **2008**, *29*, 2482–2486.

71. Yıldırım, İ.; Çiftçi, U. Monitoring of pesticide residues in peppers from Çanakkale (Turkey) public market using QuEChERS method and LC–MS/MS and GC–MS/MS detection. *Environ. Monit. Assess.* **2022**, *194*, 570. [CrossRef]

72. Kataoka, H.; Lord, H.L.; Pawliszyn, J. Applications of solid-phase microextraction in food analysis. *J. Chromatogr. A* **2000**, *880*, 35–62. [CrossRef]

73. Mills, G.A.; Walker, V. Headspace solid-phase microextraction procedures for gas chromatographic analysis of biological fluids and materials. *J. Chromatogr. A* **2000**, *902*, 267–287. [CrossRef]

74. Zheng, K.; Wu, X.; Chen, J.; Chen, J.; Lian, W.; Su, J.; Shi, L. Establishment of an LC-MS/MS Method for the Determination of 45 Pesticide Residues in Fruits and Vegetables from Fujian, China. *Molecules* **2022**, *27*, 8674. [CrossRef]

75. Farha, W.; Rahman, M.M.; Abd El-Aty, A.M.; Kim, S.W.; Jung, D.I.; Im, S.J.; Choi, J.H.; Kabir, M.H.; Lee, K.B.; Shin, H.C.; et al. Analysis of mandipropamid residual levels through systematic method optimization against the matrix complexity of sesame leaves using HPLC/UVD. *Biomed. Chromatogr.* **2016**, *30*, 990–995. [CrossRef]

76. Galani, J.H.; Houbraken, M.; Wumbei, A.; Djeugap, J.F.; Fotio, D.; Spanoghe, P. Evaluation of 99 pesticide residues in major agricultural products from the Western Highlands Zone of Cameroon using QuEChERS method extraction and LC–MS/MS and GC-ECD analyses. *Foods* **2018**, *7*, 184. [CrossRef]

77. Frenich, A.G.; Vidal, J.M.; López, T.L.; Aguado, S.C.; Salvador, I.M. Monitoring multiclass pesticide residues in fresh fruits and vegetables by liquid chromatography with tandem mass spectrometry. *J. Chromatogr. A* **2004**, *1048*, 199–206. [CrossRef]

78. Kaphalia, B.S.; Takroo, R.; Mehrotra, S.; Nigam, U.; Seth, T.D. Organochlorine pesticide residues in different Indian cereals, pulses, spices, vegetables, fruits, milk, butter, deshi ghee, and edible oils. *J. Assoc. Off. Anal. Chem.* **1990**, *73*, 509–512. [CrossRef]

79. Galván-Portillo, M.; Jiménez-Gutiérrez, C.; Torres-Sánchez, L.; López-Carrillo, L. Food consumption and adipose tissue DDT levels in Mexican women. *Cad. Saúde Pública* **2002**, *18*, 447–452. [CrossRef]

80. Kim, S.W.; Abd El-Aty, A.M.; Rahman, M.M.; Choi, J.H.; Lee, Y.J.; Ko, A.Y.; Choi, O.-J.; Jung, H.N.; Hacımüftüoğlu, A.; Shim, J.H. The effect of household processing on the decline pattern of dimethomorph in pepper fruits and leaves. *Food Control* **2015**, *50*, 118–124. [CrossRef]

81. Antonious, G.F. Residues and half-lives of pyrethrins on field-grown pepper and tomato. *J. Environ. Sci. Health Part B* **2004**, *39*, 491–503. [CrossRef]

82. Antonious, G.; Hill, R.; Ross, K.; Coolong, T. Dissipation, half-lives, and mass spectrometric identification of endosulfan isomers and the sulfate metabolite on three field-grown vegetables. *J. Environ. Sci. Health Part B* **2012**, *47*, 369–378. [CrossRef]

83. Fantke, P.; Juraske, R. Variability of pesticide dissipation half-lives in plants. *Environ. Sci. Technol.* **2013**, *47*, 3548–3562. [CrossRef]

84. FAOSTAT—Food and Agriculture Organization of the United Nations. 2017. Available online: https://www.fao.org/faostat/en/#home (accessed on 25 November 2022).

85. Liu, X.; Yang, Y.; Cui, Y.; Zhu, H.; Li, X.; Li, Z.; Zhang, K.; Hu, D. Dissipation and residue of metalaxyl and cymoxanil in pepper and soil. *Environ. Monit. Assess.* **2014**, *186*, 5307–5313. [CrossRef]

86. Lu, M.X.; Jiang, W.W.; Wang, J.L.; Jian, Q.; Shen, Y.; Liu, X.J.; Yu, X.Y. Persistence and dissipation of chlorpyrifos in brassica chinensis, lettuce, celery, asparagus lettuce, eggplant, and pepper in a greenhouse. *PLoS ONE* **2014**, *9*, e100556. [CrossRef]

87. Rahimi, S.; Talebi, K.; Torabi, E.; Naveh, V.H. The dissipation kinetics of malathion in aqueous extracts of different fruits and vegetables. *Environ. Monit. Assess.* **2015**, *187*, 1–9. [CrossRef]

88. Jung, D.I.; Farha, W.; Abd El-Aty, A.M.; Kim, S.W.; Rahman, M.; Choi, J.H.; Kabir, M.H.; Im, S.J.; Lee, Y.J.; Shim, J.H. Effects of light shading and climatic conditions on the metabolic behavior of flonicamid in red bell pepper. *Environ. Monit. Assess.* **2016**, *188*, 1–9. [CrossRef] [PubMed]

89. Zhao, E.; Xie, A.; Wang, D.; Du, X.; Liu, B.; Chen, L.; He, M.; Ye, P.; Jing, J. Residue behavior and risk assessment of pyraclostrobin and tebuconazole in peppers under different growing conditions. *Environ. Sci. Pollut. Res.* **2022**, *29*, 84096–84105. [CrossRef] [PubMed]

90. Alavanja, M.C.; Bonner, M.R. Pesticides and human cancers. *Cancer Invest.* **2005**, *23*, 700–711. [CrossRef] [PubMed]

91. Mostafalou, S.; Abdollahi, M. Pesticides and human chronic diseases: Evidences, mechanisms, and perspectives. *Toxicol. Appl. Pharmacol.* **2013**, *268*, 157–177. [CrossRef]

92. Kumari, D.; John, S. Health risk assessment of pesticide residues in fruits and vegetables from farms and markets of Western Indian Himalayan region. *Chemosphere* **2019**, *224*, 162–167. [CrossRef]

93. Bellanger, M.; Demeneix, B.; Grandjean, P.; Zoeller, R.T.; Trasande, L. Neurobehavioral deficits, diseases, and associated costs of exposure to endocrine-disrupting chemicals in the European Union. *J. Clin. Endocrinol. Metab.* **2015**, *100*, 1256–1266. [CrossRef]

94. van Wendel de Joode, B.; Mora, A.M.; Lindh, C.H.; Hernández-Bonilla, D.; Córdoba, L.; Wesseling, C.; Hoppin, J.A.; Mergler, D. Pesticide exposure and neurodevelopment in children aged 6–9 years from Talamanca, Costa Rica. *Cortex* **2016**, *85*, 137–150. [CrossRef]

MDPI

St. Alban-Anlage 66

4052 Basel

Switzerland

www.mdpi.com

Foods Editorial Office

E-mail: foods@mdpi.com

www.mdpi.com/journal/foods